Food Industry R&D

Food Industry R&D

A New Approach

Helmut Traitler

Birgit Coleman

Adam Burbidge

WILEY Blackwell

Registered Office
John Wiley & Sons, Ltd, The Atrium, Southern Gate, Chichester, West Sussex, PO19 8SQ, UK

Editorial Offices
The Atrium, Southern Gate, Chichester, West Sussex, PO19 8SQ, UK

For details of our global editorial offices, for customer services and for information about how to apply for permission to reuse the copyright material in this book please see our website at www.wiley.com/wiley-blackwell.

Library of Congress Cataloging-in-Publication data applied for

ISBN: 9781119089391

A catalogue record for this book is available from the British Library.

Wiley also publishes its books in a variety of electronic formats. Some content that appears in print may not be available in electronic books.

Cover image: Hong Li/Gettyimages

Set in 10.5/13pt Times by SPi Global, Pondicherry, India

Printed in Singapore by C.O.S. Printers Pte Ltd

10 9 8 7 6 5 4 3 2 1

Contents

About the Authors

Helmut Traitler has a PhD in Organic Chemistry from the University of Vienna–Austria. He was an Assistant Professor and Group Leader of a Research Team for Westvaco in Charleston, South Carolina, USA, working in Vienna, Austria. He joined Nestlé Research in 1981 and later became a member of the Editorial Board of the *Journal of the American Oil Chemistry Society*. At Nestlé, his roles have included Head of the Department of Food Technology; Head of the Combined Science and Technology Department; Head of Nestlé Global Confectionery Research and Development, York, United Kingdom; Director of Nestlé USA Corporate Packaging in Glendale, California; Head of Nestlé Global Packaging and Design, Nestec Ltd., in Vevey; and Vice Preseident of Innovation Partnerships at Nestec Ltd., working in Glendale, California, as well as Vevey, Switzerland. In August 2010, he cofounded Life2Years, Inc., a start-up company in the area of healthy beverages for the 50+. Helmut is the Senior Innovation Connector for Swissnex San Francisco, a public–private partnership organization sponsored by the Swiss government with offices in Beijing; Bangalore; Rio de Janeiro; Cambridge, Massachusetts; and San Francisco,. He is actively involved in technology spin-offs of mission-noncritical know-how for the Jet Propulsion Laboratory (JPL) in Pasadena. He has most recently been involved in codeveloping food products in the area of sports and is the author of more than 60, mostly peer-reviewed, scientific publications, 26 international patents, book chapters and 2 books.

Birgit Coleman is a strategic thinker and Connections Explorer in her current role at Swissnex San Francisco. Her expertise includes recipes for growth through internal innovation and external strategic partnerships with the goal of building a disruptive innovation pipeline for the clients of Swissnex San Francisco. Before Swissnex San Francisco, Birgit worked for the energy drink company Redbull North America, and IBM in Vienna, Austria—her home country. She holds a Masters Degree in Business from the University of Vienna.

Adam Burbidge obtained a BEng and later a PhD in Chemical Engineering from the University of Birmingham in the United Kingdom. Subsequently he worked as a postdoctoral research at IFP (Lyon, France) and the University of Cambridge before taking up a lectureship in Chemical Engineering at the University of Nottingham, United Kingdom. After a couple of years at Nottingham he returned to University of Birmingham as a member of the academic staff. During his time in academia he supervised a number of PhD students and ran a research group with interests in rheology and particle technology, which was funded by a combination of grants from government and industry. After several years in academia, he left this field to take up a post at the Nestlé Research Center, near Lausanne in Switzerland. At Nestlé he has headed various groups in the foods science and technology department with a general focus on applying soft matter physics approaches to food. Adam has published more than 45 research articles with over 900 citations; he reviews for more than 20 journals and several government and industrial granting agencies. He is a member of the Society of Rheology and lives in the canton of Vaud in Switzerland with his wife, two daughters, and three cats.

Foreword

The ancient Greek philosopher Heraclitus tells us: "war is the father of all things." When it comes to the history of modern industrial research and development (R&D) organizations, he is spot on. The gigantic science and engineering projects during World War II (e.g., the Manhattan Project, to name a prominent one) provided versatile models for big technical companies to organize their R&D after the war. Bringing together basic research and advanced engineering under one roof seemed the best way to concentrate the R&D efforts for novel technological developments.

The need of innovations was not the basis for those ideas, but the view that "basic research is the pacemaker of technological progress," so eloquently expressed in 1945 by MIT professor Vannevar Bush' s report "The Endless Frontier.[1] He predicted that pursuing new basic scientific concepts would lead to novel products and services.

However, the last 30 years saw the rise of the age of innovation. The wide availability of creative tools, like personal computers and the Internet, has leveled the playfield between companies. The belief that basic research alone while performed in-house will drive growth has lost its adherents. However, innovations carry risk of failure; a fact leading straight to restrictive countermeasures and the epidemic application of processes and procedures in R&D. This has generated the lamented tunnel vision of contemporary industrial R&D.

Against this background, Helmut Traitler together with his coauthors tells us his story and views about industrial food R&D. His findings are based on his personal observations, experiences, victories, and failures. Traitler does not waste his time in anecdotic nostalgia. He has crystallized general insights and new ideas from his years in R&D of Nestlé and beyond. These ideas comprise new means for a revival of creative R&D organizations. It is fascinating for me to follow his analysis having together worked on innovations in Nestlé for many years.

His critical review rightly focuses on people and structures. It is in these two areas where the unforgivable management sins are occurring. Importantly, Traitler documents that people and structures are not independent. They form a self-enforcing feedback loop where mediocrity supported by management structures stifles creativity and kills innovations.

The actionable outcomes of Traitler's analysis are collected in the second part of his book that presents "possible futures" of food R&D. He provokes the reader to change perspectives on consumer insights, external innovations (universities and other solution providers), and the future development of the food industry. He tops his analysis with "disruptive outlooks" describing new ways of organizing R&D based on testable business models. Traitler belongs to the few who make their advice personal, having it grounded in lively experience. I hope that innovation managers will heed his advice.

Heribert Watzke
Lausanne,
September 2015

[1] http://www.nsf.gov/about/history/vbush1945.htm

Preface

Research and development (R&D) not only represent a vast area of topics and heated debate but it also is a playground for much controversy of the most different kind. In academia, such controversy is often based on interpretations of data and subsequent conclusions and often debates the question of who was first to discover a particular finding and whether or not the said finding is of any value to the scientific community. R&D in corporate environments follows different rules and judgment patterns and is mostly defined and driven by costs and consumer relevant targets, or so one may believe. There is, however common ground among these two worlds: both strive to maximize knowledge, although for different reasons and in different ways. Equipment and scientific rigor may be similar or identical, however their usage, approach, and interpretation are different. This book discusses history and background of today's food industry as seen by consumers, academia, and the industry itself, and several chapters are especially dedicated to new and disruptive approaches to R&D. Is your company presently restructuring its R&D organization? I bet it is! Then this book is definitely a must-read for all professionals in the packaged goods industry as well as students who aspire to contribute to this new type of industry forcefully driven by R&D!

Acknowledgment

This was not an easy book to write. Let me explain why. During my professional life I had worked most of the time in research and development (R&D) and only shorter periods of time in other parts of my former company such as packaging operations or open innovation and partnership management. Here's the dilemma: because of my deep insight into R&D organizations of food companies I can easily see their inefficiencies and flaws; however, I also feel a deep loyalty and constructive understanding for R&D and everyone who works in this minefield of a food company and probably other companies as well. On the one hand I can understand how people in the food R&D act, and react and on the other hand I can also understand those who criticize those actions and reactions and ask for change. However, change is always expected to start elsewhere and fingers are pointed so easily.

My first thanks go to all those former colleagues in the various R&D organizations whose paths I have crossed and who have taught me everything I know today, parts of which I had the great opportunity to write down in this book. And I also thank all those unknown, competent, loyal, and creative R&D people who were and are responsible for what is happening in R&D today, good or bad, because the learning from them was tremendous.

My special thanks go to my two co-authors Birgit and Adam who at the end had regretted having encouraged me to nag them. Birgit was already an extremely capable and innovation driven co-author of my first book, so it was almost easy to convince her to become part of this endeavor too.

This being my third book on a food industry–related topic in a fairly short period of time required a lot of patience and especially understanding in my direct vicinity. A special thanks goes to my wife Thérèse; she was the one who brainstormed with me on chapter outlines and contents, and all this from an unsuspected and untainted, just pragmatic and reader-oriented position. She also had to endure my status reports and ups and downs in the progress of this book project.

My son Nik Traitler helped me design all figures, as he did for my first two books. I believe you will appreciate the simplicity and clarity of all figures. I would also like to thank my dear friend and colleague Heribert Watzke, who has been kind enough to write the foreword to this book, a fitting yet very concise introduction.

Last but not least I would like to express my sincere gratitude to my publisher, Wiley Blackwell, and the entire team behind for their continued trust in the ability of my coauthors and myself of not running out of ideas, which we believe are worth sharing with you, the readers. For this I want to send you the readers my very special thanks!

Part 1

What we have today and how we got here

1 A typical food R&D organization: Personal observations

I know that our R&D probably costs twice of what it could cost but I don't know which half to cut.
Helmut Maucher

1.1 INTRODUCTION

Let us play a game. I like playing games. Research and Development is typically abbreviated to "R&D," and that's good, because otherwise books, publications, presentations, discussions, and such would become too long, always repeating "Research and Development" instead of using the short, catchy, and dynamic sounding "R&D." The game is easy: find as many other meanings for R&D as you possibly can and list your favorite ones. Let me give you a few examples: rich and dumb, raw and delicious, real and daunting, rooster and duck, ready and done, ruined and defunct, researched and developed. Ooops! The last one is almost the same as research and development, however, there is an important difference: research and development means that everything—or almost everything—is still ahead of you, while researched and developed means: done, ticked off, executed, found, and made. I can tell you from deep and longstanding personal experience that the past tense R&D (the "Red & Ded") is the real dream of every company executive in just about any company in any area that you can imagine, while the "R&D" is a real headache for them.

Figure 1.1 illustrates our "find-other-meanings-for-R&D" game.

This book is mostly about this headache and how to heal it. It's not about "pills" that can help the headache go away but rather a change of lifestyle, or more correctly a new approach to R&D, especially in the food industry so that the headache goes away by "natural" means or doesn't even come up in the first place. This is not an easy feat, yet it is worthwhile, no, essential to undertake, otherwise R&D in the food industry will cease to exist because in case of doubt which half to cut, CEOs and executives of the food industry will simply cut it entirely, partly out of frustration and partly out of simply not knowing better. Members of the business and commercial community and even those of the manufacturing and procurement community

Food Industry R&D: A New Approach, First Edition. Helmut Traitler, Birgit Coleman and Adam Burbidge.

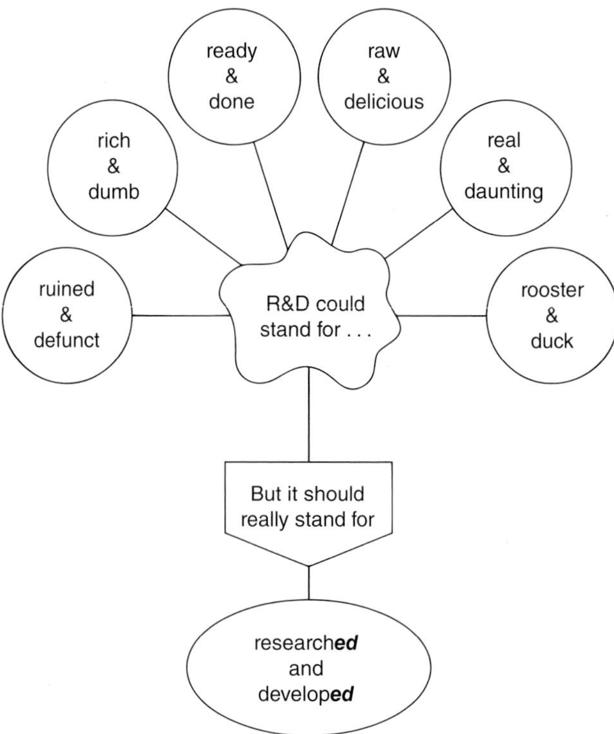

Figure 1.1 R&D could stand for ....

seem to have little understanding for anything that is R&D "tainted" and a bit more basic and difficult to understand. This is unfortunate but it's a reality, which cannot be neglected easily or even discussed away. Chapter 10 will in much detail discuss the scenario of an R&D-centric food industry organization in which scientists and engineers "call the shots" and hold the reins of the company. I can already hear business and commercial leaders shout out in unmistakable ways what they think of such a scenario. Their discontent will even be bigger when the following hypothesis will be discussed and analyzed.

1.1.1 Business people always know better

Commercial experts mostly know all about their field of action and because science is not easy to understand, let alone being learned in evening school, they don't even attempt to understand scientific and technical or engineering details. On the other hand, there are oodles of scientists and engineers who have ventured out to get an MBA degree in addition to their technical degree. What I want to say here is simply that scientists can fairly easily acquire expertise in business, whereas commercial and business people hardly ever, or better *never* go back to college and do a science degree; it's simply too time consuming and not an easy undertaking. At least that's true for the food industry. The situation is different in the pharmaceutical and even chemical industry. And it's even more so true for the finance and banking industry, although it is difficult to assume that scientists could have come up with a bigger mess than the good people in banks and the world of finance in recent years.

But let's get back to our topic. Not to let scientists and engineers get ideas that they could rise in the hierarchy, they have to be put within their boundaries and not get the idea that they could become CEO of a company. Best shot is CTO and that's that. However, the hard-to-dispute reality is that it is possible for a scientist or engineer to acquire a business degree, but it's virtually impossible and almost unheard of that business leaders acquire a science degree. There are of course exceptions to this and, again, especially the pharmaceutical industry has many examples in which medical doctors or scientists have become CEO.

Well, after this initial rant, which sets the tone of the book, quite on purpose, let's get to business and talk about R&D in the food industry, what it is, what it might evolve to, and lastly, what it really could be.

1.2 A LOOK BACK IN WONDERMENT

Corporations always had R&D departments or functions or labs or just a few "crazy" guys inventing something. The strange—or maybe not so strange—fact seems to be that most corporations in the past were founded based on a great idea and invention by a technical genius or guru who then, together with partners, mostly business savvy ones, turned this idea into some great business of sorts. History shows that most often the tech gurus who were the inventors and the real basis for the new company to exist in the first place and to grow were quickly put to the side and so-called business people, the serious guys, the guys who knew, took over. This pattern over so many years has been rather successful; rare are the exceptions that it is taken for granted today that corporations have to be led by business representatives and not the technical guys.

1.2.1 Innovation is everyone's business

This is not to say that all innovation and invention is of technical nature and only technical people can innovate; far from it. Technical innovation would not fly if it were not accompanied by business innovation. There are many important business innovations especially in logistic and supply chain, manufacturing, stock keeping, procurement, and purchasing, and even accounting and financing and new approaches to legal matters have come to pass. What I definitely do not count in this list is cost cutting. Cost cutting is probably the antithesis of progress, innovation, and sustainability. When highly paid managers don't know any further they either go "back to basics" or call for a "cost-cutting initiative" or worse, they do both. I emphasize this so much because at first sight this really is counterproductive to innovation and progress. On the other hand, restricting means and tools and making life a bit more difficult for everyone is, after all, not such a bad thing. Restriction and scarcity of available means can actually provoke and even sustain innovation. As for all things in life, the balance and especially timing are of the utmost importance to steer the ship of R&D smoothly and successfully.

My first venturing into corporate R&D dates many years back to the late 1970s and happened while I was working as assistant professor at the university in Vienna, Austria. The department in which I had worked on my thesis has had a long-standing contract with a U.S.-based pulp and paper company and was mainly interested in knowing all about lignin, this "nasty" side product that you get when you work wood, mainly pine wood, to paper and cardboard. Lignin is almost like gold in as much as it almost can't be broken down into useful chemicals and so, in these days and to a high degree even today, went into asphalt or hair dyes as additive and similar, low-added value applications. The chemical structure of lignin, a "compound" of several aromatic (six carbon atoms) rings, would make it a highly valuable candidate for many applications if it could only be broken up in meaningful and cost-effective ways. Anyhow, the company wanted to really turn lignin into something valuable and had supported, financially and with a lot of patience, ongoing research in this area in our department for many years, however, without too many striking results to say it nicely.

1.2.2 Let's go and have a drink

Management representatives of the company visited us twice a year and made the "sacrifice" to come to Vienna from somewhere in the Carolinas and meet with us, hear us out, encourage and criticize us, and mostly also to go out with us in the evening, preferably to one of the rather famous—or infamous—"Heurigen" restaurants where you drink the local wine and, if you would like to, bring your own food.

The wine is served in glasses of a quarter liter, so they are rather big; enthusiasm and the atmosphere easily carry you away to drink more than you actually can take, especially if you are not used to it. You can imagine that the mood soon became cheerful and everyone was happily complimenting everyone else for the good work, the great results, and so on. This was the first time in my professional career that I totally grasped the real meaning of "wining and dining" and especially its central importance in the corporate world.

The next day was always extremely tough and demanding. This was partly because the hangover that typically resulted partly from overindulgence (a euphemism) and partly from the tough questions that were asked during the meetings. We had to prepare reports and the running joke was that the company representatives would always expect results that were similar to "diluting water with water" and make everything even more cost efficient and ultimately cheap. In those days, writing reports meant typing them on typewriters and then duplicating them on "spirit duplicators." We had photocopy machines, however, the costs of one copied page were still rather high, so we went to this cost-efficient and ultimately cheap device of duplicator, and I vividly remember having rolled out hundreds, if not thousands, of pages for such meetings. So, on top of having come to an intimate understanding of the term wining-and-dining, I also learned the true meaning of rolling out early on in my R&D career.

1.2.3 Never give up and continue to hope

Interestingly enough, these meetings in my recollection never turned "bloody," messy, or unpleasant. We always came to good conclusions and expectations of even more promising and especially conclusive results next time. I was in charge of the research group for the better part of 3 years but the group had already existed for more than 10 years when I took over. So, looking back, I have to assume that the results were always sufficiently promising that the company continued to ask for more work to be done and more discoveries to be made, as little and insignificant as they might have appeared. This was another important early learning what R&D and especially corporate R&D was, and still is, all about: Hope. "Sister Hope" is probably the foremost driving force in R&D, even when looking back (Note: hindsight is the only exact science!) one can see that the road of discovery is filled with cobblestones of misadventures, deceptions and disappointments. But wait, there is hope beyond Hope: "Never give up" is the younger brother of the first born sister "Hope" in the world of R&D. I have to admit that these two siblings are rather weak reasons for R&D to exist, would they not have another sister, the youngest and most volatile of them all: Success. Success is what everyone wants to see and have and strives for, but little sister Success is always somewhere else and rarely with you. You always have to find her in the most unexpected places and have to keep her away from bad company and bad substances. If you invite her properly and accept her wholeheartedly, she is more likely to visit you, but again, it's not a guarantee. You have to work hard and use sister Hope and brother Never-Give-Up all the time to eventually meet Success. It'll be a sweet meeting because it will give you the ultimate justification that sister and brother were all worthwhile and most important to have. It's like with every other family: only the sticking together gets you through all difficulties and brings you to much deserved successes.

The family is of course not complete without parents like father "Finances" and mother "Business," uncles "Procurement" and "Manufacturing," and aunts "Legal" and "Marketing." And there are many more relatives, more or less distant, that you can add to this incomplete list of family members, and they are all important. And this family has to have many friends, what do I say, oodles of friends, hundreds of thousands, better even, millions of friends that look in admiration to this great family and which we can loosely define as "Consumers." This book focuses on the three siblings Hope, Never-Give-Up, and Success, in the deeper context of R&D and how these work together and give each other praise but also consolation whenever needed, especially if sister Hope has once again disappeared for quite some time.

1.3 A LOOK BACK TO THE BEGINNINGS OF A TYPICAL FOOD INDUSTRY R&D

I need to apologize to you the reader for the fact that I base my experience largely on my former company, but this is where I learned all about the typical food industry R&D organization. Given that this company is the biggest food company in the world, however, probably makes it representative enough to use it as a historic example. Moreover, throughout many years, and especially the years during which I was responsible for the open innovation program "Innovation Partnerships," I came to meet many colleagues, especially R&D colleagues from a really great number of food and food-related companies, which gave me great insight into their R&D organizations. I will, therefore, always complement this look back with examples from many other companies, large ones and also smaller ones, the so-called "SMEs" (small and medium sized enterprises).

1.3.1 It all starts with a great idea

Speaking of my former employer, the Nestlé company, I would like to remind the reader that it all started with an idea, based on an apparent and urgent need, to create a healthy and stable product; healthy for malnourished babies and stable so that it would not get spoiled. Poverty levels during these times, the mid-nineteenth century led to the situation that many pregnant mothers were malnourished or simply undernourished, and breast feeding, although by far the best and most logical way of feeding a baby was often not possible or did not give the right nutrition hence the need for complementary nutrition. And stable it had to be because the notion of a cold chain was just not around in these days and fridges were, although invented, not yet in large usage. So what was more logical than to create a dry product, in this particular case, a dry, ambient temperature stable powder through drying of the liquid formulated milk mix.

Whether you like the example or not is beside the point, but it shows two essential elements coming together, which are at the heart of every successful business, be it food or anything else, namely: recognition of a need and the idea for a technical solution and ultimately realization. Recognition of need and deployment of a solution are an unbeatable combination and should be at the heart of any business. The deployment of the solution, however, comes only at the end of an oftentimes lengthy and cumbersome R&D process with many trials, errors, and failures. At the end of the day, every R&D process always shows many more failures than successes. From the beginnings of modern corporate R&D as we know it today, in the early and mid-1950s, and ever since, this has always been a true statement: there are many more failures than successes, out of 100 attempts maybe just a few make it to the end. I have discussed this in my book *Food Industry Innovation School: How to Drive Innovation through Complex Organizations* (Traitler 2015).

Business has become smarter, or so they think by streamlining the efforts, writing detailed briefs, setting milestones, and setting out bonuses for achieved results. Fundamentally this all appears to be quite OK, except that it might kill the surprising deviations and odd turns that any project may take. I shall discuss this serious topic in much detail throughout this book; suffice to say that, the jury is still out whether such an organized approach is really hitting home and bringing many successes.

1.3.2 People were frightened

Let me return to looking back again so that the context of what is happening today and what could or should happen tomorrow becomes clearer and more understandable and believable. The beginnings of modern corporate research of my former company date back to a time in the mid-1950s, a time, which I only got to know through the crazy stories, stories of crazy and daring characters and other outlandish stories of the pioneers. The precursors of these times actually date back to the years just before World War II, but as far as I can personally judge were performed in much different ways to the ones that really begun after that war. So I will rather focus on the latter and tell the stories of what I have personally experienced plus a bit of preceding "folklore." These early years, like every other period of R&D, were influenced by societal concerns, which a few years later also found their way into the creation and findings of the "Club of Rome," for whatever they are worth (Meadows et al. 1972). These limits to growth were very much in the public discussion probably already 10 years before the Club of Rome findings and they helped to shape research directions and the resulting research projects.

The early 1960s were the heydays of the Hippie movement, which in rather straight terms left no doubt about what they wanted apart from the heavily publicized "sex, drugs, and rock 'n' roll": back to nature, simpler lifestyle, and indirectly using fewer resources. The late 1960s brought an even stronger movement, widely known as the students' "revolution" of May 1968. Many people to this day still believe that the "devil descended on earth" in those days and blame the movements of May 1968 for all evil on our planet, without a doubt. I'll let you the reader draw your own conclusions on this debate, but nevertheless, the 1968 movement brought, probably for the first time, an air of sustainability thinking: deal more carefully with finite resources. The first big so called "oil shocks" of the early and mid-1970s were yet another stepping stone to make everyone who wanted to know aware that we are sitting on finite resources and we should, after all, better be a bit more careful.

1.3.3 Are we depleting our resources?

The mid-1960s were therefore heavily influenced by this feeling of finite resources, a growing world population, and the food industry would have a special and responsible role to play in this entire context. Fear of future food shortages was almost palpable and was subsequently directly reflected in the research programs of many food companies, especially how to solve these.

The first really big research program that I heard about in my former company was initiated during these years and had one simple definition and goal: use oil (the type that comes out of the ground by drilling) as one raw material to be fed to yeast and thereby create so called "single-cell-proteins." Wow, what an idea. But hold the horses before you start ranting about the "absurdity of the idea" or anything similar. The entire idea was of course not totally new because researchers at British Petroleum (BP) had worked on the idea of feeding straight-chain hydrocarbon from their fractionation processes to yeast already as early as the mid-1950s; they called this the "proteins-from-oil-process." The real difference to the older, well-established processes of growing yeast was that instead of sugar, n-paraffins were used. BP built a first small-scale pilot plant in 1963 (Bamberg 2000). The entire single-cell-protein process became rather popular in the 1970s and even won a UNESCO Science Prize in 1976.

It has to be understood through the reading glasses of that period and then it might become a little bit better to swallow. I don't want to put today's judgment on this idea, although this process probably uses a lot less water to make the equivalent of 1 kg of vegetable proteins, let alone the freakishly high amount of water required to produce 1 kg of animal protein. But that's beside the point, and I don't want to discuss the validity of the project so much but rather the organization in a large food company. Rather surprisingly, this project quickly became the *only* project of the company's research organization during quite a number of years. This would be unheard of today. Maybe the last such "put all resources behind one goal" approach was NASA' *Apollo* mission of the 1960s. This leads me to believe that these years were more daring and higher risk-tolerant when it came to putting all the eggs into one basket, so to speak. However, I am not saying that these were "the good old days" because they were not or at least very rarely.

1.3.4 Focus, focus, focus

The positive aspects to this approach are of course the great focus, the large number of resources that were put behind one goal, and the clear goals and timelines that were defined. On the other hand, serendipity was pretty much excluded, even if many NASA scientists still to this day pretend that the invention of a ball pen that could persistently write when held upward was a great outcome of value for the public. By the way, the Soviets used pencils for that purpose; rather foolproof, isn't it? Don't get me wrong: I don't want to belittle the *Apollo* missions of these years. I personally believe that the moon landing was the greatest thing that ever happened, even better than the Beatles. I was influenced so much or rather infected by this space virus that I started to work with the NASA "Mars guys" only a few years ago and have as much pleasure today as I had then. I do admit that quite some years have passed since these early moon years and the technology of today is not only much more sophisticated but because it is so versatile, it is really useful for all of us.

But here I am deviating to the present time and even future, so let me get back to "those days," the days of the appearance of corporate R&D, especially in the food industry. One of the real reasons why a project such as single-cell proteins could not only be initiated but also run over many years—in my recollection the better part of 10 years—is most often based on personalities and personal convictions, especially those of the leaders. While on the development side of projects in the Nestlé company in those years, coffee and to some degree dairy products, including infant formula, were pretty much the main areas of R&D; the more basic research part of R&D was really preoccupied with work on single-cell proteins and this all happened because there was strong leadership influence on both research as well as development. Almost like two warriors fighting for supremacy, these two leaders were fighting for what they believed was the most important thing to do for the company. Looking back, it can easily be seen that the single-cell protein work pretty much disappeared with the disappearance of the strong leader—and believer—while projects related to areas such as coffee, milk, and infant still live to this day and are of great importance to the company because they do reflect the product portfolio quite closely.

On the other hand, even in monolithic and controlled research environments as was the case with Nestlé in the mid-1960s and 1970s, strange things could happen, and almost under radar of the mainstream, other, smaller projects could blossom. One of these projects that was initiated in 1955 (!) actually survived several decades and is believed to still exist, 60 years later, somewhere in the underground of the research project portfolio. The fun part of this is that the "really important mainstream save-the-company project" did not make it, while the unimportant small and quirky project survived for 60 years. I am not saying that this is a good thing, but it is a reality that can be found in research environments in many food companies and beyond. All these activities had—already in those days—one important common denominator, and I have mentioned this a few times: they were all organized in project structures, pretty much the same way as you would see it today:

- Goal based on a need or needs
- Description of most likely pathways to reach this goal
- Selection of most-promising technologies and resources

- Definition of milestones and timelines
- Detailed description of the end point
- Definition of closure and post analysis

You can restructure and reformulate this typical project flow as you see fit, but basically most other flows will still look similar to this one and are more often differentiated by semantics rather than substance and content.

1.3.5 A historic perspective

Let me attempt to compact this historic view to the past and present, the view on the origins and evolution of modern-day corporate food R&D in a simple descriptive beginning with the era after World War II.

The period of post—WW II between 1945 and approximately 1965 could be characterized by the drive to develop new food sources to feed populations that have strongly reduced caloric intake during the war years, which partly explains the drive for finding new sources for proteins: oil seemed to be abundant, food needed to be created by either growing (the traditional ways) or synthesizing such as in the single-cell approach.

I would suggest that this phase was followed by the post–Club of Rome period, which probably lasted from 1970 to around 1985. This period was characterized by fear of scarcity of resources, leading governments but especially nongovernmental organizations (NGOs), and in turn the ordinary consumer to take on alarmist positions. This subsequently led the food industry to create, what I would call "catch-up" type of projects or in other words, opportunistic projects.

The next, partly overlapping period was the post–moon-landing period between roughly 1980 and 1995. This period saw a lot of "everything-is-possible" attitude and the industry started quite a few rather extravagant projects, such as low-calorie fats and similar ones.

During all these years, since the early 1970s, the information technology (IT) revolution took place and really took root as something here to stay in the mid-1990s. So this led to the next important period in this historic overview.

The post-IT-revolution period, which began approximately 1995 and is still ongoing. Individualism and individual consumer-related research and development was and still is the big driver and was clever in finding health and wellness-related topics leading to many new food product propositions, such as responding to lactose intolerance, reducing salt and sugar intake, and more recently gluten-free products. Some of these are fashions and come and go, others are likely here to stay.

1.3.6 Let's cut costs

I would add one more, also overlapping period to this historic review, namely the period of the efficiency and cost-cutting revolution. To satisfy the shareholders and financial analysts when it comes to the value of each company, and that includes food companies, repeated and ongoing cost-cutting exercises have become the rule in almost every company. While it should be

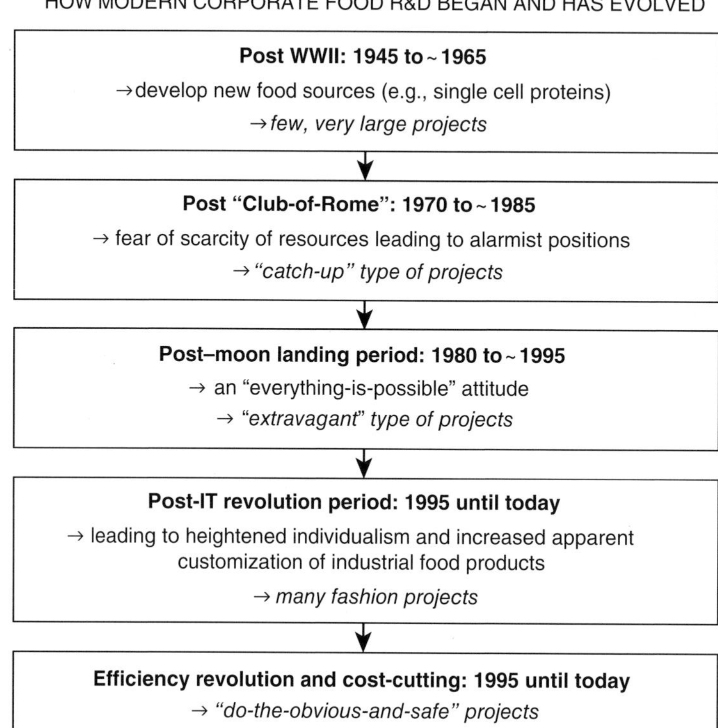

HOW MODERN CORPORATE FOOD R&D BEGAN AND HAS EVOLVED
Post WWII: 1945 to ~ 1965
→develop new food sources (e.g., single cell proteins)
→few, very large projects
Post "Club-of-Rome": 1970 to ~ 1985
→ fear of scarcity of resources leading to alarmist positions
→ "catch-up" type of projects
Post–moon landing period: 1980 to ~ 1995
→ an "everything-is-possible" attitude
→ "extravagant" type of projects
Post-IT revolution period: 1995 until today
→ leading to heightened individualism and increased apparent customization of industrial food products
→ many fashion projects
Efficiency revolution and cost-cutting: 1995 until today
→ "do-the-obvious-and-safe" projects

Figure 1.2 Historic view of evolution of today's food industry and its R&D.

obvious that being careful when spending company resources should be the norm for R&D, there is a different tone to be heard when the next cost-cutting exercise is announced, including the more-or-less exact sum as to how much should be saved. Unfortunately, this approach or rather this attitude led to the situation that only the more obvious projects, those, which have a higher promise of success even with potentially low margins at the end, are projects that are generated and run. They were often created under slightly misleading financial prospects, just to get them off the ground in the first place.

Figure 1.2 gives an overview of this discussion and analysis.

1.3.7 Food industry has simple and tangible goals

The major goals of any food company, small or large, are always to produce safe, healthy, and affordable food. And yes, it should also taste well as perceived by the consumers. It is therefore rather obvious to expect that each research organization in the food industry—and this goes for just about every industry—follows their major needs; in the case of the food industry are answers to questions related to what hides behind "safe," "healthy," "affordable," and "tasty." It is as simple and as complicated as that, and my personal take on this is that because an organization that helps find the relevant answers is rather simple in its build, people often try to make it more complex.

1.4 FROM SINGLE AND LARGE TO MULTIPLE AND COMPLEX

As a consequence of the recognition of the four major research pillars (Safe, Healthy, Affordable, and Tasty), food companies started to organize their research organizations accordingly. Departments were created that grouped logical areas together. For instance an entity that grouped macronutrients such as lipids, carbohydrates. and proteins together was named "Food Science Department." Quite logically, this department may also have comprised activities in areas of micronutrients, such as vitamins and minerals, antioxidants, and other relevant minor, active food components.

Because safety is one of the major concerns of the entire industry, an entity was founded that looked into food safety, not only from a toxicological but also from a procedural point of view: were technologies and processes in manufacturing safe and also leading to safe products?

Taste is of course the holy grail of every successful food and food product, therefore many important strides had and have to be made and work around taste and also texture and were centralized in a group typically called "Food Technology." The names may differ from company to company, and it is clear that especially smaller food enterprises had to group some of these activities together; this also happened in the large corporations. Thus, food science and food technology often became one entity, which made sense in as much as all taste and texture is a combination of ingredients—macro and micro—as well as processing and technologies. Moreover, smart selection of ingredients and optimal processes would eventually also lead to lower costs, thereby to increased affordability for the consumers.

Well, there still is the "healthy" bit. In the food industry, ever since its creation in its present-day format, healthy products were always at the forefront of every new product development or improvement of existing products. At least that's true for those companies I know and have worked with. The food industry has always looked out for help in the medical and pharmaceutical industry to find and apply metrics that could demonstrate certain health aspects of the industrially produced foods and beverages. Intuitively I would say that when George and Mildred Burr (1929, 1930) discovered, and for the first time described, the importance of essential fatty acids, we saw the onset of modern nutrition research.

1.4.1 Nutrition has growing pains

Of course there were many nutritional type studies in the years prior to these findings, however, never was there such causal proximity between a food ingredient and its function in the body described in that much detail. And it took almost 30 years until Ralph Holman and his coworkers picked up the ball and led many nutritional studies, metabolic studies to be exact, which demonstrated clearly this correlation between intake of certain food ingredients and their effect on the human body. It took the better part of another 15 to 20 years before the metabolic pathways of essential fatty acids, actually of fatty acids in general, were elucidated by the likes of Howard Sprecher (1981) and his colleagues at Ohio State University.

I use the example of essential fatty acids not only because it is fairly well documented historically but also because I have extensively worked in this area myself and have still quite a lot of affinity to this area of research (Traitler 1987).

It is clear that this is not the only example where food ingredients—macro as well as micro—meet nutritional science, and there was much work done on proteins as well as carbohydrates. However, I feel that there is greater linearity in the fatty acid research than in any other area. I can already hear protein and carbohydrate experts grind their teeth; forgive my bias toward the fatty acid arena and use this as an example of early days and still ongoing nutritional research.

In parallel to the activities of scientists and engineers in the areas of lipids, proteins, and carbohydrates, structures were built up in the food industry R&D over many years that deal with the same macronutrients from a nutritional angle. And then there are all the micronutrients such as vitamins, antioxidants, minerals, and a few others that over time were at the origin of more research groups dealing with all these additional food elements.

The next real important step was then taken once it was recognized that all these groups—food science, technology, safety, and nutrition—all required solid support systems in terms of basic understanding of pathways, interactions, and the determination and analysis of metabolic but also structural compounds. Basic science was needed, often defined by the term of *knowledge-building* or something similar.

When the company heavily worked on new coffee-roasting and extracting technologies, it was quickly seen that any process had to be accompanied by a deep understanding of the underlying chemistry, and hence the need to analyze volatile as well as nonvolatile components that had been formed, especially in the roasting process. This is just one example of the need for knowledge building to back up both, purely technical and also nutritional projects in the food industry's R&D organizations.

1.4.2 The new risk management approach: Many projects

The real message that I would like to get out here is this: in the beginning of R&D organizations we saw few, focused and fairly large projects, supported by the majority of the staff that worked in R&D and also largely supported by management who believed that this was the real way forward. The risk with this approach is rather obvious: when this one, all-encompassing project fails, there is not left to show to management and it is difficult to receive continuous funding for such an undertaking. It works in the beginning because one can always ask for more time and patience, and the results will come, after all. But when it so happens that the results never really materialize, and especially, when priorities change during a lengthy lifetime of a mega project, then you can imagine that this is the real killer and soon people would find themselves out of a job because they were experts for a specific area, not necessarily usable in a new, a different one. So, it was not at all surprising that over time, once the failure of the mega project was recognized and especially also "digested," a company—any company—would change their approach and spread the risk more evenly over more if not many projects. The unfortunate result of this is that, because there may not necessarily be more money available to support more resources, the existing resources, as expert as they may be in the required new research fields, will be spread out fairly thinly across a large number of projects. This means that focus is lost, speed of execution may suffer, and promising directions may have to be abandoned and new outcomes may just have been lost.

Nevertheless, that's exactly what happened over the years, namely the migration from few, big-and-focused projects to multiple, smaller and less-supported projects. As probably always in life, the right answer would lie somewhere in the middle and really outstanding program and project portfolios in companies would take this middle-ground approach into account. Quite naturally, in my own experience, I saw the opposite happen; we went from few and big to not only multiple yet rather *many* projects over a period of 20 or so years. As an example, an organization of 100 scientists and engineers could have as many projects and sub-projects and tasks and such with the additional burden that those who ran a project, the project leaders, would not only work on their own project but also would have to contribute to projects of their colleagues; this could sometimes mean that a scientist worked 30 percent of his or her time on the own project, 20 percent on project 2, 20 percent on project 3, 20 percent on project 4, and 10 percent on project 5. And, because everyone in a high-performance organization—whatever that is and I will discuss this later in the book—is expected to work more than 100 percent might be found to contribute 10 percent to project 6 and, why not another 10 percent to project 7.

This, of course, is a rather ridiculous situation and you can imagine that the quality of the outcome of any given project is rather doubtful, let alone the timelines, which may have shifted ever so often. This was, and is clearly an undesirable situation but was the case in the transitional years of a food R&D as I knew and still know it. The names of the game are: flexibility, versatility, and good salesmanship of the results. And everyone played that game, and to some degree, still plays. Management as well as the scientists and engineers, they are all connected in this negative spiral of doing as much as possible with as little resources as one might just get away with.

1.4.3 Too many projects? No problem, reorganize

I have personally experienced this project expansion movement and, as a reaction to this, strong efforts to again reduce the number of projects. This is a tedious, time-consuming, and costly effort, and I will discuss this in much detail in Chapter 2. Let me just say this here: scientists and engineers are typically creative people and what typically is done when management requests a reduction in projects, the first reaction is to reorganize the portfolio without really giving up anything. Similar activities or work areas are all of a sudden grouped together and new "super projects" are created. The old projects are reclassified and are for instance called "tasks." So, from initially 100 projects, the number might be reduced in almost no time to for instance 50 or even less but all of a sudden consisting of 100 tasks in total. This looks good on paper and may calm down management's excitement to have a more streamlined R&D organization and project portfolio but at the end of the day nothing has really changed.

Let me, however, introduce the following. There are two main reasons why an industry, and especially the food industry, work on many projects simultaneously. First, you cover much ground of potential relevance and importance and thereby spread the risk and secondly, scientists and engineers can better "hide behind" a large number of projects. One of the projects I am working on or participate in will hopefully eventually succeed, even if many of my other activities might just not. Chances of being associated with a winner are bigger.

However, the food industry is not alone in this and I could experience similar approaches in the chemical industry when I had a chance to work with some of the big ones in Germany. It is my assumption that the reasons for having a portfolio with a large number of projects are always the same as just mentioned in every R&D organization. Again, I shall discuss this in much detail in the following chapter and throughout the book.

1.5 WHY DOES THE FOOD INDUSTRY NEED R&D AFTER ALL?

For some, this may appear to almost be a blasphemous question, a question that cuts right to the heart of the matter, and which should maybe not even be asked, less even discussed. There is R&D, "always was there," so why ask this question. It almost sounds like the question as to why we have marketing or why we have a moon.

1.5.1 Million dollar answers to the million dollar question

Well, one simple answer is of course: because they can. It is probably not a good answer but certainly a valid one. Another answer could be because from a tax point of view R&D activities can be seen as a cost, although they really are an investment, and therefore represent a nice tax-deductible chunk in the balance sheet.

Yet another answer could simply be because the origin of almost every company and especially a food company, as shrouded as it may be in the long distant past, was always linked to a tangible idea that came out of an R&D like brain, setup, group, or just a few people working together and concocting something of potential value for the consumer and thereby for the company as well.

And here is another reason: R&D activities always look good and their outcome can be measured (e.g., in the value and strength of intellectual property, that is, number of patents, product launches as a consequence of R&D activities, and similar considerations). These metrics can be used by the financial analysts and can have a positive impact when a company's value is calculated with direct impact on its share value. Yet another reason for the existence of the R&D organization in a company might be to increase the company's standing in the eyes of the consumers. Just imagine a food company that sells their products without any credible and reasonable R&D activities behind these products. Another reason could simply be that management believes that a good and successful R&D shines back on them and gives them some glory.

There might be other logical or less logical reasons for the existence of R&D such as tax advantages or regulatory requirements or even others, which I have not listed here. Let me just mention the one reason, which I believe is the ultimate one, why any food company, small or large, locally, regionally, or globally active, traditional or more progressive has an R&D organization at the heart of their company. It is simple and one can probably best see this in the "SMEs" because it is so obvious there and not hidden behind sophistication and excessive and abundant resources: without an idea, based on consumer-needs recognition and its technical elaboration to render it "manufacturable," no food industry, and no other industry for that matter, would exist, thrive, and grow but would ultimately cease to exist.

1.5.2 Here we go: Justifications

Let me discuss the various reasons that I have listed in some more detail and find out, how much, if at all, these reasons contribute to the overall answer to the question: "why R&D" and how we can use these arguments to strengthen the position of R&D in your company internal discussions. And from personal experience I do know that you can use every argument you can get to not necessarily justify R&D in some of the tough meetings where this question comes up but to put you in a more comfortable and relaxed position, because no such justification is really needed.

My first answer was: "Because they can." Yes, it is true that often things are done because one can. It is as simple as this and, surprisingly enough, is often a really strong driving force in the decision-making process. You may want to recognize that, based on my personal observations, the element of because we can always underlies any type of decision and subsequent consequences.

There are many examples of this approach in every industry's R&D, although it is most often a function of "affluence" of the company. The smaller the company and the lower the budget that can be set aside for R&D, the more pragmatic the approach to this and other reasons such as need for innovation or consumer expectations prevail. With increasing size of a company, the overall budget for R&D in absolute numbers goes up and this attitude becomes stronger and stronger. I have seen many projects that were mainly done just because of this reason: we can, therefore we do. It sounds strange but reality is that this happens more often than one would think. And I have to say that it's not always a negative thing because the success rate of R&D activities that were based on this approach is not really lower than for streamlined and focused ones. So, an R&D organization that exists and operates because the company, its management, and its decision makers justify its existence on because we can is not necessarily less efficient and successful than any other that exists for any of the other reasons that I have briefly mentioned and which I shall discuss in more detail.

1.5.3 Because we can is a great reason!

Let me give a few examples of successful R&D organizations—successful defined as successful consumer products developed by this organization—that largely existed and still exist on the premise "because we can." Note that this is my personal judgment and you may not be in sync with me on this.

When Nespresso® was created as far back as the mid-1980s, it was mainly a small group of less than a handful of mostly technical experts, typical R&D people, who had an idea in which they strongly believed and which they had the strongest conviction to realize and bring to market. They had to operate almost in hiding; one could call this skunk work and management knew about it but did not want to hear about it. From an organizational standpoint,

Nespresso® survived these first years because of the attitude of because we can. If other reasons for the existence of R&D that I have briefly mentioned had come to play, such as good for the standing of the company or its management, great in the eyes of the financial analysts, or even good for tax write off, Nespresso® as a company might not have seen the success it ultimately could achieve.

This has of course also to do with patience. It is patience that a larger, more affluent organization can have because it is invested in so many, diverse, and already successful ventures. Smaller companies have a much tougher time in this "waiting for success," and it is another example for my observation that the attitude of because we can is mostly found in larger and more affluent companies. Large companies have the means to wait for success, although not all large companies do this. This has a lot to do with the financial valuation of any company; it is like with credit ratings: the more open credits you have, the more late payments you have made, more litigations you may have had, and the lower your credit score. Same with corporations: the lower the number of apparent product launches and successes, the larger the number of projects, especially long-lasting ones, the more goodwill is invested into ventures and collaborations, the worse the note that financial analysts give to your company.

Let me come to the next reason that I have briefly mentioned: taxes or rather the optimization of these. Although financial support of an organization such as R&D is actually an investment into the present and especially the future, for tax reasons it is simply considered a cost to be deducted from profits, thereby bringing taxable revenue down a notch. R&D in the food industry is typically anywhere between 0.5 to approximately 2 percent, and such money can have an important impact on the balance sheet. I do not want to emphasize too much on some companies' situations where rather complex tax constructs justify royalty payments and where it is not only good to have an R&D organization—because they can or because it reduces taxes—yet because other royalty payments would not be justifiable anymore. An example for an R&D organization that largely exists for these reasons is the large and global Nestlé organization. They are not alone in this, and other companies have a similar approach. I am not suggesting that tax considerations are the only reason why their R&D organizations exist, but they contribute to some pretty important degree.

1.5.4 New product development is everything, or is it not?

Another justification for the existence of R&D organizations is the straightforward, functional one: because it is necessary and it is at the heart of any new product development and innovation, especially in the first phase of any new product and process development. This justification is probably the most logical and best understood one; however, from personal experience, R&D organizations would not exist if that was the only reason for them to be. So, it is only one complementary part of the entire puzzle.

The consumer electronics industry and the automotive industry also have an important part in this because it is in industries like these that this justification for the existence of their R&D is, in my assumption, the strongest. Of course, every industry will boast that it is innovative and puts a lot of emphasis on new product development and wants to demonstrate their cutting edge side to their consumers. Food industry is no different from this type of approach, however, it is often more of a lip service than a reality. I absolutely do not intend to suggest that innovation is not the driving force, yet it does not always take the important stand inside a company. Again, this functional reason of necessity for new product development and innovation is a complementary element and partial answer to the question of why an industry, why especially the food industry, needs R&D organizations.

The next reason on my list was the desire and need to satisfy financial analysts in their quest to evaluate and valuate companies. Financial analysts, like all economists, like to measure and use certain metrics to feed algorithms and come up with valuation numbers. There is nothing basically wrong with it, but one has to especially differentiate between types of industry. Large chemical companies, I mean the really large, global ones, may apply for more than 1000 patents per year; some automotive companies, the ones that are active in innovation, may have numbers in the hundreds of patents per year; while a large food company may apply for anything between 100 and 250 patents per year. Despite the fact that patents are only one element in the total IP portfolio of any company, their number is an easy metric to understand and use for valuation calculations. There are several other IP aspects such as manufacturing secrets, trade secrets, recipes, and marketing plans that all need to be taken into account when evaluating strength of a company's market position and its value. Patents are typically created by R&D and so are, to some degree, manufacturing plans. Hence, R&D has an important justification for existence when looking with the eyes of financial analysis and evaluation of any company.

1.5.5 Consumer is king

Another important reason for R&D to exist is to show consumers that the company is serious about its products, from all aspects (e.g., development of new, improvement of existing, better safety, healthier ingredients, better taste, less packaging, and ultimately better price-to-quality ratio). This aspect is not to be underestimated because, in my own experience, is maybe one of the most important justifications of all. After all, the consumer is king, or rather queen, and if a company would decide to neglect this truism it would quickly disappear from the marketplace. Most or all of these aspects are performed and realized by R&D, and therefore R&D has a nice sweet spot in any company. This sweet spot does, of course, not come for free, and expectations are typically higher than Mount Everest or Mauna Kea for that matter. But that's what comes with the sweet spot: the urgent need to deliver the innovative new products and processes as fast as possible.

Let me come to the final reason that I had mentioned at the beginning of this section: top management looks good if they can say that their R&D is great, top, world class, or whatever other attribute they might find for praise to the outside world or inside the company. The inside praise is rarer and always happens when R&D had come with something good and valuable for the company. I have seen many times how quickly top management's mood can turn depending on the internal reporting of successes or failures. First of all, the wind blows into the faces of individuals and entire sections; departments can be wiped out by the scorn of top management, which can hurt the well-functioning and organization of entire R&D groups. Luckily this does

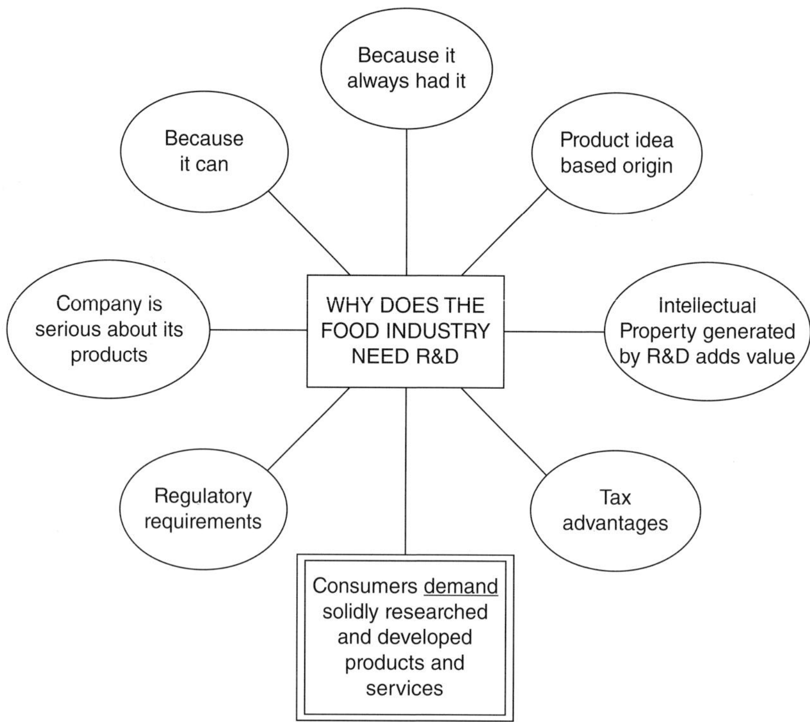

Figure 1.3 Reasons why R&D in the food industry exists.

not happen too often because R&D people are typically smart and they normally come up with successes or nothing. "Nothing" is nothing to report, therefore most or all of what is reported is successes. So, you see, no harm done. Or is there? Well, I shall discuss this in some more detail later in this book. However, I have seen top managers lose patience because of receiving "success reports" that were too easy to see through in terms of what the real quality of success was and subsequently putting research money somewhere else.

Figure 1.3 illustrates many or most of the reasons why there is an R&D organization in every food company.

1.5.6 It's all about long-term thinking, stupid

To terminate this section of this first chapter, let me say one thing, which I believe answers the question of "why does the food industry need R&D?" best. In hindsight, and hindsight is known to be the only exact science, the great ideas and solutions for new and innovative product and process development always came from the R&D organization, small or large, food or other industries, local or global. And because we do not want to disturb this fine balance of creative minds and curiosity too much, we let R&D be R&D and come with the next great thing soon. There is, however, one unifying reason for R&D to exist in any type of corporation, and which was not mentioned yet; it is the patience to wait out results combined with the culture of long-term thinking. I do not want to embark on a definition of what is long-term at this point because it is most certainly defined differently by different players in the industry. In conclusion, this is not to say that today's R&D organizations should also be tomorrow's, and that is exactly the topic that I shall discuss, analyze, and synthesize throughout this entire book.

1.6 SUMMARY AND MAJOR LEARNING

- In the introduction I suggested to play a little game: find different meanings for the acronym R&D. I suggested that the ultimate meaning of R&D is not "Research & Development" but "researched & developed" (i.e., executed and successfully done)!
- R&D: headache versus solution. R&D's role is not always clear and easy to understand and always requires justification, or so it seems.
- Scientists and engineers versus business or rather with business: there is not always a clear answer; collaboration and respect are key.
- Looking back to the early years of corporate R&D in the industry: much paper was created, hope prevailed.
- The book focuses on the three major drivers for R&D in any industry: Hope, Never-Give-Up, Success.
- The early days of the food industry R&D: focus on resources and feared scarcity defined research programs.
- Large and monolithic projects dominated the playfield (e.g., single-cell proteins, coffee technologies).
- During the evolutionary years of food industry R&D we came from few and large projects to many and small projects.
- This led to a strong dilution of expert resources over many not so significant projects, however helping to minimize risk of failure by this large spread of activities.
- It was suggested that modern food and nutritional science started with the discovery of the role of essential fatty acids in the late 1920s.
- Project themes during the evolutionary food R&D period evolved around ingredients and processes, first macronutrients such as lipids, proteins, and carbohydrates followed by all other minor ingredients such as vitamins, minerals, colorants, texture-building agents, surface-active ingredients, all coupled with progress in food processing technologies.
- The existential question "why does the food industry need R&D?" was asked and discussed in some detail.
- Major reasons are: because they can, tax optimization, functional reasons (new product development), financial considerations (to please the analysts), improved standing with consumers, and top management looks good.
- The unifying reason, however, is the willingness of good companies to wait for success and long-term thinking.

REFERENCES

Bamberg, JH, 2000, *British Petroleum and global oil, 1950–1975: The challenge of nationalism. Vol 3 of British Petroleum and Global Oil 1950–1975: The challenge of nationalism*, JH Bamberg British Petroleum Series, Cambridge University Press, Cambridge, United Kingdom.

Burr, G, Burr M 1929, A new deficiency disease produced by the rigid exclusion of fat from the diet, *J. Biol. Chem.*, vol. 82, pp. 345–367.

Burr, G, Burr M 1930. On the nature and role of the fatty acids essential in nutrition, *J. Biol. Chem.*, vol. 86, pp. 587–621

Meadows, DH, Meadows, DL, Randers, J, & Behrens, WW (1972), *The limits to growth*, Universe Books, New York.

Sprecher, H, 1981, Biochemistry of essential fatty acids, in *Prog. Lipid Res.*, vol. 20, pp. 13–22.

Traitler, H 2015, *Food industry innovation school: How to drive innovation through complex organizations*, Wiley-Blackwell, Hoboken, NJ.

Traitler, H 1987, Recent advances in capillary gas chromatography applied to lipid analysis, *Prog. Lipid Res.*, vol. 26 no. 4, pp. 257–280.

2 A typical food R&D organization: The world consists of projects

The real reason why an R&D organization has so many projects is very simple: people can better hide behind all these projects.

Hans Wille

2.1 ALL R&D WORK IS PROJECT BASED

The title of this section is certainly not a surprise to you. There is hardly a business organization of any kind that does not speak of projects and that defines all work done inside the organization as being a project. It might not come as a surprise that not all projects are equal, and there are many different approaches as to how projects and its individual subparts are defined and dealt with.

What is a project? I am not talking about suburban housing projects in New York and elsewhere, but well-defined and resourced sections of work that happen in many if not all companies in similar ways.

"In contemporary business and science a project is defined as a collaborative enterprise, involving research or design, that is carefully planned to achieve a particular aim" (Oxford English Dictionary, nd). Projects can be further defined as temporary rather than permanent social systems or work systems that are constituted by teams within or across organizations to accomplish particular tasks under time constraints (Manning, 2010). An ongoing project is usually called (*or evolves into*) a program."

2.1.1 Project has many meanings

Looking more closely into these definitions, one finds a couple of important key words that are always more or less loosely used in the context of projects. Here's the list:

- Collaborative (enterprise) or collaboration
- Involves research or design (I would interpret this as "by design")
- Carefully planned

Food Industry R&D: A New Approach, First Edition. Helmut Traitler, Birgit Coleman and Adam Burbidge.

PROJECTS . . .
. . . are result driven
. . . have time constraints
. . . go across organizations
. . . involve research and design
. . . are collaborative
. . . progress along milestones
. . . are carefully planned
. . . are finite
Content
Time

Figure 2.1 Meanings of "project".

- Aiming to achieve a particular aim (typically called "goal," along defined "milestones")
- Projects are temporary (or finite) and are not permanent systems (social or work)
- Projects go across organizations (unfortunately they not always follow this)
- They need to accomplish particular tasks (I interpret this as "milestones" and "final results")
- Projects happen under time constraints (either self-inflicted or by market or social pressure)

This list is quite a handful, and it is probably most worthwhile to go through this in some detail and analyze and discuss the meaning and value of these different key words.

Let's begin with *collaboration*. After the word *communication*, it is probably the second-most used and heard word in any corporate environment. Good communication or lack thereof is one of the great miracle words that can fix or fail just about everything. In similar fashion, this is also true for collaboration, or lack thereof. It is one of these words that almost can act like a magical wand and that is largely used in any fitting or not so fitting situation. It is all about collaboration; you need other people, other experts, to successfully run and complete a project, a task, or any type of work. So we hear and yet, this is not always true or even necessary. I will use the word *teamwork* in this context of collaboration, as much of collaboration boils down to teamwork.

Figure 2.1 illustrates the various meanings of *project* based on the two important elements of each project: content and time.

2.1.2 Third-generation R&D

Let me reflect back to a popular book of the 1990s: *Third Generation R&D: Managing the Link to Corporate Strategy*, which had the spirit of teamwork written all over the book. Here's a short description of the book from Amazon's site:

Written by three senior consultants from Arthur D. Little, this book provides managers with a new approach that will make R&D a truly competitive weapon. Relates how R&D management has evolved from the naive "strategy of hope" approach of the 1950s and 1960s, when companies spent lavishly in the vague expectation that something good would result, to the more systematic approach of the past two decades. The third generation of R&D is a pragmatic method for linking R&D to long-term business planning. It shows managers how to: integrate technology and research capabilities with overall management and strategy; break down organizational barriers that isolate R&D from the rest of the company; foster a spirit of partnership and trust between R&D and other units; and create managed portfolios of R&D projects that match corporate goals.

In my most recent book, *Food Industry Innovation School: How to Drive Innovation through Complex Organizations*" (Traitler 2015) I have written the following on teamwork:

My definition of teamwork is...a loose community of like minded people who have different, yet complementary expertise and who find themselves together because they are passionate about a common idea. Where I make a sharp distinction to the more sheep-like (pardon my French) description above under what I believe it is not, is this collection of like minded people who find together because of the passion for the cause and not because management told them so. It's almost a self-forming entity with great conviction and strength, less manipulable but also less controllable. This is probably bad news for management but it's great news (GREAT NEWS) for the success of your innovation project, but of course also for your company and after all for the consumer in the end (pp. 230–231).

I shall discuss the role and nature of teamwork in some more detail in this chapter.

2.1.3 Strategic business units became popular

I do remember that the book was one of those must-read publications that was used in company internal management seminars and was heavily used in many meetings and discussions. Linking R&D activities to defined strategies was an important recognition during the early 1990s. This was also the time, that based on findings and recommendations of consultants, the "McKinseyization" of companies took place, especially of my former company, which manifested itself in the creation of the so-called strategic business units (SBUs). This probably made a lot of sense in those days, and it made sense for large corporations to organize themselves by clearly defined and almost-independent business units. I use the word *almost* because, in hindsight, it became clear that there was a strong and clear-cut interdependence between SBUs and the central company organization. I don't want to embark on describing the structural details of the types of SBUs that I became familiar with in those years, but the field of operation for these units was rather small. One has to understand that a food company, especially a large and global one, has a tremendous portfolio of local, regional, and some global and strategic products. The "some" makes all the difference because these global and strategic products were and still are to this day the ones that fall under the responsibility of an SBU.

Depending on the type of products, be it coffee, chocolate, ice cream, culinary products, or dairy products, the split between local and regional on the one hand and global and strategic on the other could mean that there were only 25 to 30 percent of strategic products in the portfolio of the SBU; all other products, although in the same group of food products, were still under the responsibility of other managing units, typically the individual markets where these products were distributed and sold. By definition, SBUs only have a say in the strategic product portfolio, hence the name. From the onset of this structure, this was a serious limitation and was always brushed over by top management, who could not admit that there were serious flaws in this construct. It is clear that it is difficult to optimally organize a large consumer goods company that sells billions of individual product units per day in one-fits-all optimal fashion but the construct of SBUs was probably one of the weaker solutions. Instead of giving a business that is linked to a product group, for instance chocolate and confectionery, the full responsibility for all its products for all geographic locations around the world in which their products were manufactured, distributed, marketed, and sold, this solution was only a hybrid one and not satisfactory.

2.1.4 Organization is everything

Anyhow, such structures were implemented in most large corporations around the globe and everyone had to live with it, whether they liked it or not. There was a time, when a company like Cargill had more than 80 (!) business units. Nestlé has less than 10 and today has a mixed bag of business units and so called independent, stand-alone businesses.

Smaller corporations, which are not operating globally but rather locally or regionally, have an easier life in this respect; their business units typically have full responsibility for the product group they are representing and the line of command is simpler and easier to follow.

The bad news for R&D in such an SBU construct is the fact that it typically gets caught in the middle between business needs of strategic nature requested and briefed by the SBUs; of more regional nature, requested and typically briefed by a market organization; and their own ideas that they believe are of consumer value but have found no sponsor yet. It's like constantly being between a rock and a hard place with the additional uncertainty of not knowing when it will become better. Yet, all work is project based and projects have to be aligned to the strategy, they have to be briefed, they have to be sponsored, they have to have the necessary resources money and people-wise; and they have to follow certain pathways, rules, milestones, deliverables, and ultimately goals and end points. Lots of things to consider if you think that it's mainly an idea with value for the consumer. In all this organization, spontaneity falls through the cracks or is only dared by those who have the guts to look the other way.

2.1.5 Freeze the project design

Let me continue with the topic of collaboration and teamwork and the structured approach to this via briefed and sponsored projects. This approach appears to be almost too much of a constraint to build around innovative project work that should have a certain amount of freedom so that it can succeed with delivering the best outcome possible, even if it is not the desired or hoped-for one. This brings me to the next keywords that defines *project*, namely "research and design and research by design." I have already mentioned that there is a lot of planning that goes into every project and much thought is given as to what and how, why and when, for whom, and also for how much? There is certainly a lot of designing of a great project involved in this process, so much that you even hear R&D people talk about "design freeze." This means that they have achieved such a detailed degree of planning and preparation that it should not be changed anymore, otherwise costs and time would just run away. Well, here again, spontaneity gets lost and promising sideways (i.e., picking up promising leads that do not necessarily lead to the originally planned outcome) are overlooked and not pursued. Even if the individual researcher might have a hunch that this could be a promising direction, the brief from the sponsor does not allow for this. How to get out of this vicious cycle of need for briefs and sponsors and need for great results has become the daily bread of every person who works in an R&D organization.

2.1.6 How free can you be?

There are certainly different approaches and solutions to this to be found in different companies, and some handle this more freely than others. One of the solutions is to create the possibility and freedom for each individual researcher to set a certain amount of time, typically 10 or 20 percent of their work time, aside and work on new, own ideas, and which are not necessarily briefed. This sounds like a great way out, but is, in my experience, not used as such for simple fear of not having enough time to work on the briefed and sponsored ideas. I will discuss this aspect of finding the right work balance between briefed and own ideas and how to succeed in this difficult "game" in more detail later in the book. It also depends on the individual in the R&D organization as well as the company culture. Typically, the larger the organization, the more structured and "stiff" the project atmosphere might be, while in smaller companies, I typically found a greater willingness to take risks and support projects outside mainstream briefed projects. The most prominent example of this situation I experienced in packaging companies, typically packaging converters. Larger converters were typically following projects related to large, mostly stock orders, the mass products, and spent relatively little time on working on custom made, bespoke solutions for more personalized and unique packages. When visiting smaller converters I would regularly see an impressive lineup of innovative new packaging prototypes based on different materials, exciting shapes, and attractive graphic solutions. So, one conclusion of this observation could be that there is an inverse relationship between size and desire to innovate. The smaller the company, the less there is to lose; so, let's go innovate. The larger the company, the more seems to be a stake and everyone becomes careful and does not want to take real risks.

2.1.7 Small is beautiful

I do admit that this conclusion is a rather general one but fits, in my experience, most of the situations. It is a topic that I will discuss again, especially also how to escape this seemingly asphyxiating paradigm of small=innovative versus large=risk averse and non innovative. But again, I have seen this so many times that I was starting to believe that there is an unfortunate truth behind this paradigm. I say unfortunate, because I believe that this paradigm goes far beyond packaging and could potentially be true for R&D organizations in general: the larger, the less innovative; the smaller the smarter. I do not believe for a second that this is really a general rule that also applies to R&D as a whole, but there is something to it. I have seen large R&D groups that were lucky enough to have many highly talented scientists and engineers whose main job seemed to be to compete for projects or even perform similar research in different groups just to compete with each other, mostly following safe bets and pathways and not being willing to take risks.

On the other hand, I have been able to collaborate with smaller teams and see them in action, taking risks, and achieving great, exciting, and profitable results. There could be a simple solution to this apparent problem, namely to break up large organizations that resemble a supertanker into a fleet of small and agile boats that take the risk to find the right winds and sail into directions that the large supertanker would not dare to pursue. I shall take this topic up in later chapters of this book and analyze it in Chapter 10.

So, if small was really the solution to becoming innovative also in an R&D environment, why isn't every company, every R&D organization, striving to break large and difficult-to-steer and navigate groups into smaller, more independent units? There is no easy answer to this question, and probably, if there was, it would be more than just one. Let me speculate a bit here, albeit the fact that my speculation could be described as educated guesses. The first and foremost reason that comes to my mind is simply that every change is always seen as a threat, and people and especially organizations are hesitant when it comes to change of any kind, be it people or structures. Moreover, the larger an organization, and the more numerous the people that work in such an organization, the more difficult if not outright impossible it will become to change anything. Even small modifications and adjustments are seen not only as a threat but also as a sign of mistrust in the wisdom of the current leadership as well as the hierarchical levels below.

2.1.8 Pipelines

A term that you often hear in the context of projects and project related work is *pipeline*. There used to be a time when *rocket ship* was popular, but today it is *pipeline*. Another term that is often used is *innovation funnel*. The funny thing about all these terms is that they describe closed systems: the rocket ship is totally closed, although burned fuel is flowing out at its end; the funnel is a restriction device; and the pipeline is simply a tube with boundaries and something (oil, water, apparently projects) flows through the pipeline. By virtue of natural laws of physics and flow dynamics, it is expected that, under the assumption that there are no leaks, the same amount that was filled into such a pipeline at the front end (the beginning of the project pipeline) will ultimately come out at the back end, the "intermediary end." I say intermediary, because that's by no means the end-end of the project, but most often this is the hand-over moment to other players in your company, outside the typical R&D environment.

Confused? Well maybe only slightly. But let's look into this in a bit more detail. The basic idea of the project pipeline was to build in hurdles, rather check-points, which are based on strategy, tactical feasibility, economical factors, resources, and capabilities as well as timing and, quite obviously, consumers' expectations. I don't intend to ridicule the term *pipeline*, but it is an example of corporate speak that was invented by people who have probably not done their homework and was blindly accepted by all other ones in the company. A pipeline can only "lose" inefficient projects if there was a leak; actually there are some consultants in the industry who show such pipelines as porous membranes that let the inefficient and not-so promising projects literally seep out. Anyhow, it's not too important how you visualize this, and if your company works with the imagery and terminology of *pipeline*, you just have to go with it and use it as best as you can. Just make sure that everyone in your company has the same understanding regarding back end and front end of the pipeline and that there is an overall understanding that there are more projects going in at the front end than will eventually come out at the back end, which then becomes the front end for further, most likely business-related project work.

2.1.9 Try it out first

There is yet another important point I want to analyze and discuss here: preliminary exploration and conceptualization. These two elements are of potentially tremendous value as, by definition, they are short and come to the point quickly. By "to the point" I mean the moment when a sensible and fact-based decision can be taken, whether the concept is of value and should be pursued or whether it is of low or no value and should already be stopped at this point in time. Based on experience and observation, the time frame needed for preliminary exploration and conceptualization could be as short as 12 weeks and should not take more than 6 months. Typically the practical time frame lies somewhere in between these two extremes.

Preliminary exploration simply means to test one's idea, one's product or process proposal, one's underlying project idea in simple and realistic ways. This involves some preliminary technical as well as business tests, nothing deep, many assumptions still yet as close to a consumer reality as possible. This should also mean to involve different resources with complementary backgrounds already in such an early phase and let them together hash out a preliminary result that can already be translated into a preliminary, tentative concept in the shape of a prototype, product concept, packaging solution, and a preliminary business plan that contains numbers and general assumptions such as: target consumer (group), geography, purpose, and reason for new product/process/service, expected costs of these (raw materials, un-negotiated at this point in time), manufacturing costs, and multiplication factor to translate to a consumer price. The latter is typically in the order of two to three times ex-factory costs, sometimes even higher. Ultimately you have to dare and make assumptions about expected sales (revenue) and possible profits. You have to balance all this off to the overall expected development costs—rather investments—and apply reason, logic, and especially sound judgment.

A concept that comes in the form of a prototype supported by a preliminary yet already substantially thought through business plan is probably the best starting point—or killing point for that matter—for every successful project and should be applied to any new project proposal, actually even in project review situations of ongoing projects. I am aware that many companies prescribe such an approach to their R&D community, however, the fact that most or almost all requests for new project developments come from business units or businesses within the company, it is believed that the brief that is sent out to the R&D group is sufficient and a full-fledged project may be started at this point in time with little or no preliminary testing and concept building. That is a huge mistake and should be corrected in every company, small or large, as quickly and efficiently as possible. This brings me to the next important topic in this chapter, managing projects in meaningful and highly efficient ways.

2.2 PROJECT MANAGEMENT

I have this strange idea that somewhere deep in the shroud of history of any company, when everyone seemingly did everything and there was little specialization yet, someone in such a company said to his or her colleague: "you should probably manage this project and take responsibility for it, otherwise we are not going to get anywhere." Maybe the word *manage* was not used because man had not come of age yet and it is likely that words like *lead* or *look after* were used. To this day, manage and lead or "project manager" as well as "project leader" are still widely used, often confounded without really being able to distinguish the two different meanings. To be totally honest with you, I have never really been able to see the difference; however difference must be as both terms are used in the industry and there are often even related job titles of individuals. Let me give it a try though and I would kindly ask you, the reader to pick up at the end of the paragraph and add your own, personal wisdom, influenced by your company's culture and organization. One tends to believe that manager is more important than leader and therefore there could be a hierarchical difference. Maybe there is, although I don't really buy it. Who would you rather follow: a manager or a leader? I know the answer for my part: leaders have to have charisma, personality, and a strong belief in the goals as well as knowledge how to get there. A manager is a manager is a manager and should know how to manage. Period.

2.2.1 Manage or lead? Manage and lead

I do realize that I oversimplify, even pose leading questions and giving tainted answers. However, please show me the flaw in my analysis and I shall willingly change this point of view. But let's not get hung up with this, my point is a different one: it's the one about managing a project by leading people. It's as simple and as complicated as this. Because the central question: "what is good project management" wasn't asked yet and therefore has not been answered. Let me give it a try. I say good project management and don't want to embark on terms such as *great*, *efficient*, *outstanding*, or *exciting* but am happy with good and especially solid and sustainable project management. Sustainable is important because you want to be sure that the outcome of the project that you have successfully managed is a continued success and, almost more importantly, you are called in again to manage another project. I was tempted to say: "run a project" but let's stick to manage, at least for the moment.

Good and solid project management with you as the project manager starts with a smart selection of your project, credible content and expectations and, last but not least, the choice of the best people that you can find and involve in your project work. So, in a first summary, we have three elements to consider:

- Smart selection of project, mainly based on…
- Credible content and expectations, as well as…
- Right group of people to work with you on the project. Note that the group may consist of internal as well as external resources.

As I suggested, the origins of project management are not really clear but, in my experience, the evolution of project management was mostly driven by administrators and upper management, both of which wanted to have an easier grip on any project that happened in their company. The motivation for administrators was simply to mold any project work in reigning admin rules, while the upper management always was and still is striving for the infamous "30,000 feet view." Even if the latter is rather meaningless it makes for good corporate lingo. Yes, I do understand that upper management has limited amount of time on their hands and therefore really want to only hear the shortest and possibly most striking about your project. What this also means, in my eyes, good project management always contains a good portion of most appropriate and adapted project reporting. At the end of the day, the most efficient project management is worth nothing if the results cannot be told in the right manner.

2.2.2 Select the right project and deliver

I don't want to make this section a discourse on good project management, however, it is important to realize that project management is one of the cornerstones of any project success and therefore has to be taken seriously. As briefly mentioned, the three elements (i.e., project selection, credible content, and expectations as well as the right resources) make or break your project. There is more, however. Duration of the project and proper timing is such an additional element. You can work on the greatest project, manage it superbly, and have all the right resources and people with the right attitudes however, if you cannot deliver on time all this was in vain. Therefore, good and efficient project management also means being able to take shortcuts, to smartly go off track to get to the goal faster and, maybe most importantly, accepting non-perfection. The latter is a frequent point of contention and breaks many otherwise great project managers: they strove for perfection and "died in beauty." Great project, unfortunately it took too long and, inconveniently exceeded the agreed-on budget. Tough luck. Yet, it has got nothing to do with luck, it ultimately is a sign of poor project management and the wrong behavior of the project manager.

So let me add three more elements that are an important part of good project management:

- Duration and timing of project
- Being able to take shortcuts for faster delivery
- Accept a nonperfect outcome thus respecting timelines and budget

2.2.3 Teamwork is not everything, it's the only thing!

In a previous section, I briefly discussed the notion and importance of teamwork that seems to be at the basis of every successful project, or is it not? What I mean by that is simply the fact that every member of a team has his or her distinctive role to play and that, despite important joint efforts, the individual and his or her contribution can still be distinguished, very much what you hear and experience in a musical performance by a philharmonic orchestra but also by the funkiest rock band: the effort and virtuosity of the individual player contributes to the overall result yet can still be recognized. That is teamwork at its best. When looking at traditional jazz the teamwork happens slightly differently: first and most often are performed in a joint session, followed by individual players highlighting their capabilities, terminating again jointly. It's a worthwhile thought to assume that teamwork in a research and development-centric project could be carried out in such a fashion: the project team performs together, jointly, in unison at the onset of the project, it then ventures out to show the strengths of the individuals, and at the end comes together again as a team to perform the final act of the project.

It's not really important in which way the teamwork is applied to any project, as long as the individual contribution is not sacrificed in the name of teamwork itself. This point also lies at the center of debates about remuneration of individuals versus teams and oftentimes makes it hard to close to impossible for any manager to judged individual performances of the team members because they were so totally diluted in the overall final outcome of the project. That's what makes discussions between manager and team member often not only difficult but also outright unpleasant as the individual's contribution to the overall success is often unclear. To resolve this apparent dilemma it is really advisable to acknowledge the individuals' roles in every project review meeting that happens during the lifetime of any project, thus avoiding ambiguity and wrong judgments as much as possible.

So, I can say that there are three more elements that are of importance when running and managing a project:

- Teamwork in the format of bringing different and complementary resources together into one project
- Despite working as a team, clear distinction of every team member's contributions, which must be highlighted in project review meetings thus making performance measurement an easier and more just task for management.

Figure 2.2 shows the nine important elements necessary for good and efficient project management.

2.3 ALL PROJECTS ARE SPONSORED

Nothing's for free: neither lunch nor a project. That goes even more so for projects in the industry and especially in the food industry. In the first section of this chapter, I have introduced the history and story of SBUs that came into existence in the early 1990s in most companies. This was a result of almost-epidemic proportions and was caused by a "business virus" that

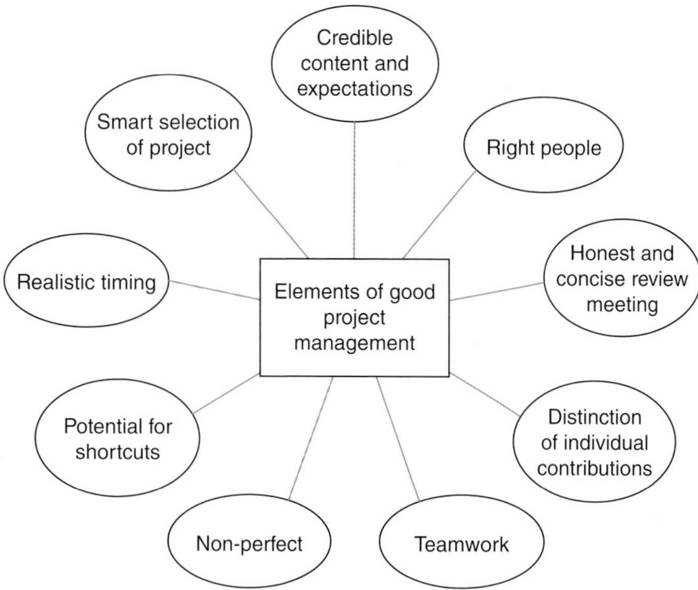

Figure 2.2 Elements of good project management.

I would call "McK-Businessvirus 1." Fundamentally there was nothing wrong with this because the industry was in dire need of a new and efficient and strongly business-based internal functionality. This not only brought new structures with new faces and new roles but also brought a new approach to how a company would see and use their R&D organization. Before the existence of SBUs, the relationship between the business part of a company and the R&D part were rather serendipitous to say the least. They were often unstructured and based more on personal relationships between people from the two sides and thereby often less efficient than they could have been. There was much duplication, often purposefully supported to enhance internal competition between R&D groups that worked on similar topics.

2.3.1 SBUs: The new, old kid on the block, happy anniversary!

This is not to say that all was bad and there were not great R&D breakthroughs during these times. On the contrary, there were many. There was, however, the realization that such breakthroughs did not come often enough and that the ratio between cost and size of the prize was too high. "Something's gotta give," or better, something had to change. Together with the improved controllability of R&D that came with the introduction of SBU structures, the efficiency of the R&D output as well as its cost efficiency could simultaneously be improved. However, there was a fine line that business had to balance to give R&D enough necessary freedom of their actions yet steer them in ways that were good for business.

Today, my former company could (if they wanted to) commemorate the twenty-fifth anniversary of the coming to life of the SBUs and an organization that gave new direction and meaning to the coexistence of business and R&D, coexistence, that all of a sudden became collaboration!

That was, and in part still is, probably the greatest achievement that can be related to this then new organizational renewal. In the meantime, much has become routine and the reality of the ultimate decision game caught up with all those involved in business units. This might be handled slightly different in different companies but the reality I know best is simply: SBUs, loyal to their name, have two basic jobs: first to define and help deploy a companywide-accepted strategy for their segment of business (for instance "chocolate and confectionery") and second, "look after" the portfolio of products with strategic importance (meaning strategic importance to the business in question but also the company at large) and make sure that both new product development as well as all other downstream activities are performed in the most appropriate and margin improving ways. Sounds like quite a handful, especially when you know that typically such business units have no real budget; they are accountable for a large portion of the budget linked to their business but have no real money of their own. The limited role and power of SBUs was already briefly mentioned previously.

2.3.2 Accountability and responsibility: A "repartition" of roles

When business units request R&D to perform new product development, and now we come back to R&D, then the bill for this service is ultimately paid by the R&D community and not the SBU. So, the SBU requests and R&D executes and pays. Although the SBU is accountable for the outcome, R&D is ultimately responsible for how well and efficiently the money was spent depending on the achieved results. It is unlike marketing, where the marketing manager has a more or less substantial budget and directly decides if, how, and how much of this budget is going to be spent on marketing campaigns and initiatives ultimately designed by the marketing person or team. Accountability and responsibility is under one roof, making this approach a more efficient one.

When the Nestlé company stepped up to the task to better structure its R&D organization in the mid-1990s to create what was then called, and still is, a conglomerate of "product technology centers" (PTCs), much discussion began exactly defining the exact individual roles of both organizations: the SBU and the PTC, which had its major activity in the corresponding product group. Because the headquarters of the Nestlé company as well as a rather large portion of its R&D activities are located in French-speaking parts of Europe (Switzerland and France), a new technical term to describe the activity of defining who does what was created *repartition*, faithfully grounded in the new-speak of "Franglais."

Anyhow, lengthy discussions were followed by even lengthier meetings and exponentially lengthy paperwork, the so-called "repartition documents." These documents were almost contracts, spelling out the exact roles, responsibilities, and accountabilities of the individual players, namely R&D (the PTCs) and corresponding strategic business units. I was personally involved in such an exercise in the area of chocolate and confectionery (or "confections" as it's called in the United States).

In hindsight it was probably a good exercise, because, for the first time, clear roles were hashed out and defined, and individuals as well as organizations committed to these. However, over time, in the day-to-day grind, much of this was forgotten and bad habits such as encroaching on each others' territories became more and more the case, ultimately leading to a situation where, seen through my eyes, confusion started to reign and indecisiveness won over the let's-do-it attitude. Much had to do with the biggest flaws of the SBU approach, namely that the budget was—and is—not in their hands and they constantly would have to delegate budget decisions to other groups, especially the R&D organization. That is not to say that SBUs do not discuss and decide budgets that are in their product group responsibility; they do but it's almost like the proverbial eunuch in a harem, look but don't touch! In my eyes this always gave them a feeling of being powerless, which often led SBUs to secondary topics, which were less debated and needed less or no money to support. Again, all this is based on personal observation and it may well be that those who work in today's business unit cosmos have a different view and opinion. Would be nice to hear back from them!

2.3.3 SBU demands, R&D delivers

This system of interaction between business and R&D as an example defined in the way I described, is not exactly the same across all companies in the food industry, but it holds true for larger food companies. Small- and medium-sized companies often cannot afford the complexity of such structures and therefore have a direct and straightforward link between the business, its marketing and sales, and an R&D group, typically a product development group that looks mostly into renovations of their existing portfolio, although one can find product developments with clear innovative character as well. Things happen faster and in simpler and shorter ways in smaller companies, although some formalism is required too. This brings me to the type of communication that typically happens between the business, typically a business unit, and its development arm, typically an R&D group. While in the days before the creation of SBUs, requests for new product development, be it renovation or innovation, were often strictly verbal, with little written communication and were usually based on gut feeling of the requester. If the requester is really good at what he or she is doing, this may not be such a bad method, however, who guarantees that this is always the case? There was much serendipity, luck, and inefficiency built in this approach and how SBUs operate in this area are certainly more efficient, at least as far as goals and expectations of the results of any request are concerned. In the early days of the SBUs' existence, these requests, briefs as they are called, were most often meticulously prepared with much insight and consumer-based information. This obviously took and still takes time and preparations of briefs can take months to prepare.

There was much talk to formulate so called "preliminary briefs," which should only cover the preliminary, exploratory phase of a project, however, even this watered-down version of a full-fledged brief would still require substantial insight and preparation, so it was not really worthwhile to go this route. It was rather pushed on the R&D community to write their own preliminary brief, thus pushing both, accountability and responsibility for this first phase of the project entirely to the R&D community. You can probably feel from this discussion that despite the sophisticated "repartition" contracts between business units and R&D not all was entirely clear and much was still left to the initiative of individuals who got things defined, supported, and moving forward. After all, this is not such a bad thing because it can speed up the process of new product development—innovation and renovation—quite substantially.

2.3.4 A brief comes from above

By the way, this has become pretty much a habit over the many years of co-existence between SBUs and R&D, namely to be pragmatic about briefing projects; more often than not, the SBU sends a formal, thought- through request, and in other instances, R&D is so convinced that they are up to something that they write their own brief, share it with the SBU and get tacit agreement. There are, however, other ways as to how briefs of any kind come to the R&D community. This can be simply through pressure from the boss or peers. A boss or a colleague came up with an intriguing idea and you were fascinated or simply taken by it and, to please the boss through "anticipatory obedience," you just jump on the idea and make it a project. People do this more often than not and hope to not only find a breakthrough project but also make their boss happy and outcompete colleagues around them.

The other pathway is directly from higher echelons in the company, with the CEO as a top contender as potential idea giver. Just imagine that you heard the CEO express an idea that he (or she) believes no solution has been found yet and that would be totally worthwhile for the company; people feel seduced to jump on this and create a project. It's a risk and sometimes it can really work out (I had personal positive experience in this), and on other occasions, it might end up as a total flop. In the latter case you will quickly find out that you are alone with your failure. To come back to the topic of this section, namely that all projects are sponsored, let me modify this statement slightly: all projects should be briefed. However, and that is a *big* however, the real fun only starts when you can do certain things without having been told what exactly it should be and when and how and for whom it should deliver. The real fun begins when you are at the helm of your project, including its inception and justification, definition and resourcing, timing and expectations, and ultimately its great outcome. I will discuss this in later chapters. Suffice it to say that it calls for different people (or the same but with changed or modified "guts") as well as different environment.

Figure 2.3 shows a high-level comparison of the roles of a SBU and R&D.

2.4 THE PREDICTABLE ORGANIZATION

One of the consequences of the McKinseyization of companies was certainly the fact that they became predictable. Projects and related actions and activities almost became rituals, strictly to be followed by everyone involved. Rituals are not such a bad thing; however, they make for routine and one expects "absolution" at the end, provided one has followed the strict rules and layout of the ritual. So, what is the ritual? It begins, like it should, at the beginning and most companies, small or large, would say that the consumer is the starting point. They also will say that the consumers are also the end point. To me this sounds like a wormhole through a time warp, but then, let's not get off track.

2.4.1 First ritual: Research the consumer

The company, especially its business representatives, are supposed to know all about "the consumer" and his or her deepest desires, may they be outspoken and met or un-outspoken and unmet. This is quite a handful and a rather tall order. Nobody can really "know" all this, you

INTERACTION BETWEEN SBU AND R&D

SBU	R&D
Gathers consumer understanding and insight	Defines and perfects relevant technical skills
Develops strategies for the specific business	Assures strategy relevant expert resources
Formulates briefs based on insight and strategy in collaboration with R&D	Organizes and sets up appropriate project teams
Sends brief to R&D and requests execution	Delivers results according to plan

Figure 2.3 Typical roles of SBU and R&D.

can "assume," "guess," "believe," and as a consequence of this, support to gather facts and resulting directions is sought through consumer research. The latter is a definite part of the ritual and marketing typically wouldn't go forward with anything without supporting consumer research. Like with any such research, the outcome can be heavily influenced, even biased, by the types of questions that are asked in such a consumer research. I am not suggesting that bias is purposefully introduced into the equation but then I am not sure, based on many years of personal observation. Sometimes it felt like a particular marketing person or group wanted to steer the research in a particular direction because that was the favored one, thus leading to a biased and not necessary realistic result.

You may argue, and rightly so, that every market research costs money, typically a lot of it and why on earth would marketing want to spend money on something that they already know beforehand in which direction they want to go. Well, the answer is simple, there are actually two related answers: first, I can show to the company, my boss, my peers, and my colleagues that I have done due diligence and secondly, if the project turns out to go south, I can always argue that market research has guided me in this particular direction, and therefore, it's not my fault, at least not entirely.

2.4.2 From "business scenario" to "business plan"

Once market and consumer research are in, assuming that the quality of the research is the best I can have, a business scenario can be constructed, which quite obviously has to fit into the overall company or product group strategy. I purposefully use the word *scenario* and not *plan*, because it might be premature at this point in time to have a full-fledged business plan at hand at such an early moment. However, if a solid and realistic business plan can be formulated, then by all means it should be done. This will make any resulting project proposal so much stronger, even assuming that it would only serve the preliminary part and conceptualization phase of the project. It's always a good thing to be ready for what will come and not chase after the facts. Maybe I should add one word regarding the distinction between *scenario* and *plan*. The scenario goes into much less detail when it comes to the manufacturing and roll-out phase of the product or service that results from the project in question; however, it needs to have really good assumptions when it comes to defining the target consumer group, the benefits of the product or service, the target market or markets, and ultimately a good hint as to the "size of the prize," in other words, how much do I expect to earn with the results of my project, most likely after a time lapse of typically three years.

As already insinuated, the plan would also contain a lot more details regarding manufacturing and all roll-out related topics such as especially communication, marketing, and sales. However, a good and solid business plan starts with a fairly large number of more or less solid assumptions, much based on the experience of the marketing team and historic data from the business. There are many individual elements that go into any good business plan and the formulation of any such plan is labor intensive and resource consuming. It might therefore be not such a bad idea to begin smaller (with a scenario, or general concept, if you prefer) and fill in the grid with elements and preliminary findings and data as one goes along in the project work, especially the results from the preliminary exploration and conceptualization phase.

2.4.3 More rituals

I used the term *ritual* to describe the general actions that surround any project and related project work. There are so many more than just the ones that are connected to the consumer research phase of any project. Once the project gets to the R&D community, the real rituals only seem to begin now.

These are written project proposals that define everything, beginning with the title of the project, its general description especially linked to expected benefits for the defined consumer groups in the well-defined geography (the market), required resources in terms of people (the experts, internal and external), as well as money (the budget), timelines, deliverables, milestones, hurdles (and how to overcome them) and, last but not least proposing the dates and frequency of project milestone review meetings. Yes, again more meetings. I may even have forgotten some rituals or you, the reader may list additional elements of rituals that you experience and live in your company. But it's quite a handful already and makes for much administration around the actual project work.

I do not pretend that all this is bad but it smells like a lot of control and especially also self-limitation to the actual project work. Reading through this list of elements and actions, it almost looks like there is only little time left over for the practical work. This is of course not really the case; however, many technical people I know and I have spoken to in R&D organizations, and not only in my former company, perceive all this as a rather heavy burden, seemingly distracting from the "real work." Let me briefly discuss the role and value of so-called project milestone meetings. At first sight, they seem to be a real pain in the neck; unfortunately, at second sight they still are. There is however value to such meetings, provided that they would really do what they are supposed to do: give a brief update on the progress of the project, especially what was achieved, what potential snags were encountered and how you propose to repair and go forward. It's really showing the decision makers that you are on track. Basically, this can be done in 5 to 10 minutes and, if all is clear, there should not even be a discussion, unless you propose to change the direction of the project or want to terminate the project because you could not meet any or a majority of the agreed-on milestones and you do not intend to drag the project any farther than necessary. This could take some more time.

2.4.4 Projects never seem to die

By the way, in my many years of project work I may have encountered maybe two projects that were officially put to rest in such a meeting, any I may have even blown up the number to make it look more realistic. From my experience, this (almost) never happens. Admission of defeat is largely seen as what it appears to be, defeat, and defeat is seen as a stigma, negatively tainting one's career. Anyhow, project review meetings seem to be a necessary evil, which, if well and correctly done, should actually add value to the entire process. So, one should make sure to stick to the ground rules of such meetings as much as possible:

- Be factual.
- Be short and concise.
- Always refer to the agreed-on milestones and measure project progress against these.
- Tell the truth as difficult as it may be.
- Never go overboard with the number of slides that you use as visual support.
- Don't be defensive.
- Make sure you get a decision (or decisions) at the end of the meeting.

I have sat in many such meetings and I have always experienced that hardly any of the preceding points was really seriously applied; meetings tend to drag on and on and often, no real decisions are taken at the end of the review meeting.

I have mentioned *milestones* in the context of the review meeting, suggesting that one is always dealing with a "milestone review meeting," meaning that the real topics that are reviewed are the ones linked to the initially agreed-on milestones of any project. This is an important realization because many times all kinds of points are discussed in these meetings, which are not linked to the set milestones. It's a real shame because this often leads to an unsatisfactory end result of the meeting without any conclusive decisions taken. This is an oft-repeated reality and seems to be part of the ritual world of project reviews in the food industry R&D world. My suspicion is, having talked with many colleagues from other industries, that this situation I not unique to the food industry.

2.4.5 It's all about results

The most important part of any project is a tangible result; it's not about the fact that you have nicely adhered to the budget, respected timelines to the hour, did not ask for additional resources, or that you have hit all or almost all milestones; it's all about the tangible, usable, and profitable result at the end of the project. Too often I have sat in "wonderful" project review meetings where everything sounded just great, except the results were not conclusive and more work had to be done and more time is needed, and so on. It's probably part of the nature of the "beast" called "project" that it never seems to have a real end in a real result, at least at the level of R&D. This is ultimately one of the most important factors that influences the opinion of most everyone in your company outside R&D in highly negative ways; R&D always has to chase after a good reputation of being a solid partner, which delivers not only on time and on budget but which also ultimately delivers profitable results. At the end, it's all about the results; yes, it's about money and timing, too; again, you can have the best timing and the most cost-efficient project, but it's worth nothing without a great outcome.

There is one other point that I will discuss here, namely the art of discovering and realizing promising shortcuts and sidelines to your main project. This is often seen in ultimately successful projects and the result of a deviation to the originally set out and agreed-on plan have brought many great results in the project activities of the R&D group in the food industry. I do not have historic data to support this but I do know of many small examples, personally experienced or seen in projects of my colleagues. This seems to be the second nature of this "beast" called an R&D project that it is often highly unpredictable, even with the best plan; deviations and shortcuts happen all the time, and they do this rather successfully.

Figure 2.4 illustrates in an overview typical business and R&D project related rituals.

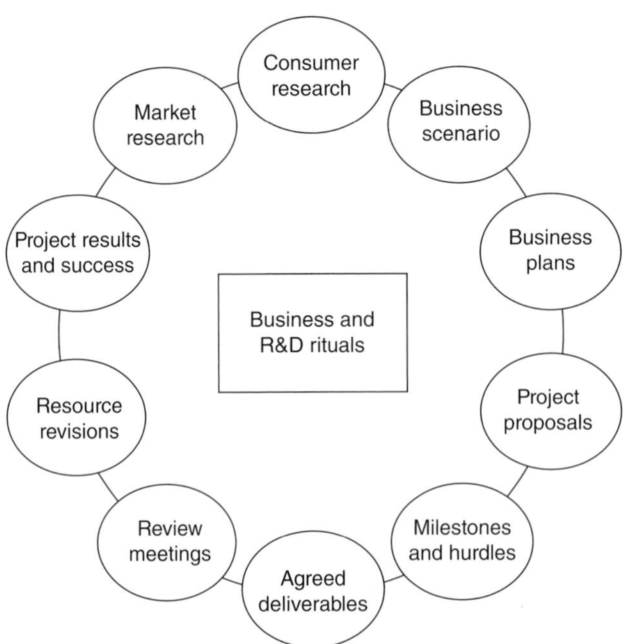

Figure 2.4 Rituals from the business and R&D.

2.5 VALUATION OF PROJECTS

A successful end to a project with a profitable result at its end can only be declared really successful if it can be measured in one way or another. This is not an easy undertaking at all, it's not trivial and has given much cause for debate, dispute, and dissonance in the industry, especially in the food industry. How do you measure the value of a project or its end result and was it all worth it after all?

In *The Food Industry Innovation School* (Traitler 2015), I have covered the topic of "Success measured" in an entire chapter and concluded that there are many types of metrics that are typically applied in the industry, some of them are what I would call "hard metrics," others, not negligible ones, are the rather "soft metrics," which are largely based on human interactions and emotions.

2.5.1 Your project could have delivered more!

There is probably no one agreed-on method and way forward on how to exactly value the outcome of an R&D project, and I have observed that those industry representatives who make such valuations rather tend to use soft metrics for such valuations, because hard metrics are so much more difficult to apply. The main reason for this is the time delay between starting point of a project, its end point, and especially the moment when the results are implemented and start to earn money. However, these are not the only points of contention. Other points are related to the amount or rather percentage of the individual contributions by each of the players along the chain of events, including all other players, such as procurement, manufacturing, logistics, marketing, communications, sales and probably a few others. Each of these has a share in the overall end result and success and, rightly so, claims it.

I have lived the following apparent contradictory situations in many project valuations: the spending that R&D is typically allowed amounts to 0.5 percent to approximately 2 percent of the revenue of a food company. It is known that the chemical industry probably spends two to three times this amount for their R&D, while the pharmaceutical industry spends a multiple of that amount, anything between 10 and 20 percent.

I cannot really speak to industries outside this type of consumer-goods industries but know for a fact that electronics or household appliances as well as the automotive industry and even the entertainment industry spend more on R&D-related activities than the food industry.

2.5.2 That's what others invest

Figure 2.5 shows an overview of R&D spend as a percentage of sales of different companies for 2011.

2.5.3 Sell your project better: Start by explaining it so that everyone can understand it

This overview clearly shows that all companies listed in this figure have substantially higher R&D spending than the food industry. So let me get back to the food industry "anomaly." While

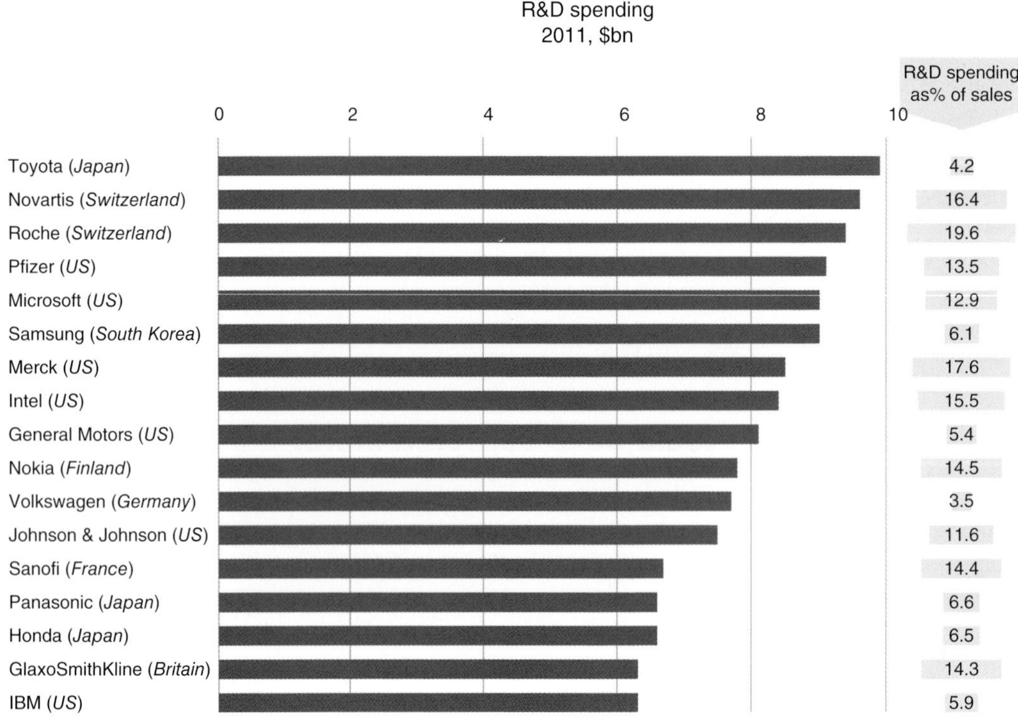

Figure 2.5 R&D spending of selected list of companies as percentage of sales for 2011. Source: The Economist.com, R&D Spending, Focus, October 30th 2012, retrieved from http://www.economist.com/blogs/graphicdetail/2012/10/focus-7

a typical spending amount for R&D in that industry is for a large company such as Nestlé in the order of 1.5 percent (some say 2 percent), and although it is lower to much lower in most other food companies, the expectations for what I would call business enhancement of margin improvement is much higher than the 1.5 percent. It is typically expected that R&D contributions have a multiplication factor of for instance four to five or contribute at least 10 percent in the overall company effort toward margin improvement. You can, probably rightly so, argue that this is a realistic expectation and there is nothing profoundly wrong with this. However, when it comes to the final valuation of the contributions of successful R&D projects to margin improvement, the argument goes more toward overall dilution because of the fairly high number of unsuccessful projects in the overall R&D portfolio. This may sound like an excuse for R&D's inability to be a better seller of success and related success stories and this is probably linked to a simple and basic fact: representatives from R&D are trained in scientific and engineering matters and terms, marketing and sales and many other company branches have representatives that are trained in communication. Unfortunately this is a reality and the main suggestion that I would make here is for R&D scientists and engineers to train and become much better in communication matters and in emphasizing the selling of understandable and credible success stories in much better ways. This is not an easy undertaking and I have extensively written about this topic and especially the value of success stories in *The Food Industry Innovation School* (Traitler 2015).

In today's R&D organizations in companies of the food industry it is, however a reality that agreed-on investments—seen as costs by the rest of the organization—are rather out of sync and proportion with the multiplication factor of such investments and the expectations linked to this. Hence, bad blood between the R&D community and other parts of the company can often be the case principally based on the out-of-sync expectations. In my eyes, in the present setup, there is no relief to this in sight or even possible. Chapter 10 will take up this point again and propose, analyze, and discuss different ways how a "futuristic" R&D organization could or rather should look like.

2.5.4 Communication is king!

In defense of the R&D community, it has to be said that communication often is not the forte of other branches of a food company either, otherwise new innovative products or renovated relaunches of existing products should have much better success with the consumers. I have mentioned the role and value of consumer research, and I just want to add that the business representatives of a company, especially marketing and sales, most of the time take the results of any given consumer research for real, especially when it comes to the quantitative research and erects a whole "numbers building" based on such research, which everyone else in the company seems to take for granted. This is especially the case, when the financial experts of the same company are called in to make all the calculations and even present their findings and proposals in large strategy and project meetings.

In principle, there is nothing wrong with such an approach, provided that everyone involved in this understands that such financial projections may be as much dreams and illusions as projected outcomes of technically driven R&D projects. This, however, is not the case and therefore makes it even more difficult for R&D to survive sanely in such an environment, especially given the chance to perform and deliver as they see best fit. Yet another excuse, you may say. I don't think so, it just reflects a reality and calls for the R&D representatives to become smarter and act in similar ways as the business community of their company; believe in what you say and do and deliver what you have said and done!

Let me close this chapter with two more short reflections linked to the valuation of projects. The first one is linked to the consumer research: everyone in a company, in your family, among your friends and acquaintances is a consumer, too; you are a consumer, too, but seem to forget this. You have an opinion and therefore you can use this admittedly personal consumer network to conduct your own, personal consumer study. It's by far better than the "palet-in-chief" or the "absolute taste" of a high-ranking executive of your company such as the CTO. You can use the gathered information to make your project even stronger, and I know for a fact and from many observations that this happens today in successful projects, almost like a best practice.

2.5.5 Speed is everything

The second and final thought in this section on project valuation is linked to the element of "time to market" that every project in R&D is strongly and intimately confronted with. Time to market or speed to market or whatever it may be called in your company is one of the most highly neglected elements in any R&D project in the food industry. I don't say this lightly but with much conviction based on many years of observation and personal experience. If you begin to think that the result of your project could be in the marketplace just meager 6 months earlier, and therefore start to earn substantial amounts of money 6 months earlier, and, on top of all this, liberate expert R&D resources 6 months earlier to begin to work on new projects, which, in turn, would again pick up speed, and additionally to these first 6 months, terminate the next project 6 months earlier..., you get the picture. I don't want to develop this as a mathematical series because it wouldn't make sense, however the message is clear by now. The speed-to-market element should really become high priority in your project work. Again, it's not taken into account enough in today's food industry R&D world and direly needs attention and correction.

2.6 SUMMARY AND MAJOR LEARNING

- All R&D-related work is project based. While the word *project* appears to be clear to everyone, I discussed implicit elements of project in more detail. I mentioned that projects are mostly collaborative, involve research, preliminary and throughout the life span of the project, are typically carefully planned, have an objective and target, and are finite, or should always be. Projects go across organizations, progress along milestones, and mostly happen under time and resource constraints.
- An important milestone in the evolution of R&D and especially R&D in the food industry was the publication of a book titled *Third Generation R&D: Managing the Link to Corporate Strategy*. The book really emphasized the importance of strategy for corporate R&D.
- Teamwork became the name of the game, this, however, made it more difficult to evaluate individual performances of members of such teams. This potentially led and still leads to inefficiencies in such teams.
- The early 1990s saw the creation of the so-called strategic business units (SBUs) in most corporations, of course also in the food industry. This brought a new organizational dimension to the structure and work of corporate R&D.
- The implementation of SBUs also led to diminishing degrees of freedom for the world of R&D, which was not always seen as a positive result.
- An important observation was discussed in relation to size of corporations and especially size of their R&D units: the larger the size, the more difficult to find real forward-looking, innovative work and results. Breaking up of large R&D organizations to smaller, more independent and agile units would be a solution to this dilemma.
- Project work in the food industry R&D is highly structured. Reasons for this are better control by management and potential resource efficiency. Downsides of this reality are a potentially diminished efficiency and innovation output. Pipelines, front-end and back-end innovation, are typically used in the industry and can be replaced by similar fitting or not so fitting terms.

- It was discussed that preliminary exploration and first conceptualization of project ideas are an extremely valuable and efficient vehicle to very quickly discard ideas that will not work.
- The role of project management as well as the distinction between project manager and project leader were analyzed and discussed. Good project management always will take the following into account: smart project selection, projects filled with credible content and realistic expectations, the right group of people from internal as well as external resources, realistic assumptions regarding project duration, recognizing valuable short-cuts and sidelines, accepting non-perfect solutions, respectful teamwork with distinct recognition of contributions from the individual team members, and honest and concise project reviews.
- All projects in the food industry's R&D are sponsored, mostly by SBUs. SBUs coexist with R&D and have project accountability, without directly paying for the costs of the overall project. These costs are borne by R&D. The specific roles of SBUs and R&D were established and were and are mostly respected. It could be said that: "SBU demands and R&D delivers."
- All sponsored projects are so-called *briefed projects*. Briefs typically contain all business relevant information that is necessary for the technical R&D group to deliver appropriate solutions, worked out along agreed on time lines and resource needs.
- Through increased structuring and detailed organization of companies, these companies become more predictable. This may lead to the belief that innovation also becomes predictable, an assumption that is not totally unrealistic. Parts of such predictability are: consumer research, market research (qualitative as well as quantitative), establishment of detailed business concepts, which in-turn lead to full-fledged business plans (all still based on assumptions!)
- Once a project is established it goes through a series of "rituals" such as the next round of detailed project proposals, project meetings, budget revision meetings, project reviews, and similar nonproductive activities.
- It was also briefly discussed that once established project, despite all good wishes and promises and all the control mechanisms just mentioned, never seem to die but take on a life of their own and can become fairly old. Some of such projects can even become very old, old such as in "several decades." After all, it's all about results.
- Valuation of R&D projects is often a point of contention between the R&D and the business community of a company. There is not really much hope to resolve this issue once and for all because there is often irrationality prevailing. Expectations by the business are much higher than the percentage spending into R&D.
- The best and most efficient way to achieve proper evaluation of one's project is linked to professional communication and storytelling about the achievements and great results. Part of good communication is the ability to speak honestly about results, including the all-important respecting of agreed-on budgets, resources, and timelines. The most striking argument in a good success story is having gotten to market faster than planned.

REFERENCES

Amazon.com (nd), "Summary of *Third Generation R & D: Managing the Link to Corporate Strategy*," retrieved from http://www.amazon.com/Third-Generation-Managing-Corporate-Strategy/dp/0875842526/ref=sr_1_2?ie=UTF8&qid=1463764133&sr=8-2&keywords=third+generation+R%26D

Manning, S 2010, "Embedding projects in multiple contexts: A structuration perspective," retrieved from http://papers.ssrn.com/sol3/papers.cfm?abstract_id=1582680.

Oxford English Dictionary, "Project," retrieved from https://www.oxforddictionaries.com/definition/english/project.

Traitler, H 2015, *Food industry innovation school: How to drive innovation through complex organizations*, Wiley-Blackwell, Hoboken, NJ.

3 A critical view of today's R&D organization in the food industry: Structures and people

I believe in innovation and that the way you get innovation is you fund research and you learn the basic facts.

Bill Gates

3.1 A TYPICAL SETUP OF A FOOD R&D ORGANIZATION

Chapters 1 and 2 described and discussed the history and basic functionality of the food industry R&D and its relationship to the business side of a company; this chapter, however, will describe, analyze, and discuss structures and people of typical food industry R&D groups. I shall attempt something that may not possible, to describe the setup and structure of a typically food industry R&D organization. In this chapter I will largely focus on the setup itself and the type of people such a setup requires and also attracts. Historically, most food industry R&D groups are organized and structured by the major food ingredients as well as supporting technologies. That sounds simple but at the end, it is not so simple after all. Why not? Mainly because the supporting technologies are of variable and changing character and follow technology trends and new findings that typically take place elsewhere. But that's not all: even the so-called food macronutrients such as lipids, carbohydrates, and proteins undergo "popularity changes" driven by public perception and frequent new nutritional and public health findings and fashions, and are thus reflected in the R&D setup of any food company. Some food companies follow fashions or real findings faster; others take more time, depending on the degree of maturity of the company and, probably more importantly, their strategy. It is a typical reaction of the larger corporations to be slower in this and often even perceive real new nutritional and health breakthroughs as potential fad and adopting the approach of wait and see as part of their overall strategy.

Food Industry R&D: A New Approach, First Edition. Helmut Traitler, Birgit Coleman and Adam Burbidge.

3.1.1 New idea? Let's wait

Smaller food companies, which typically have less of a heavy and long-standing evolutionary baggage on their virtual shoulders, act much faster and adapt to new trends more easily. This is not to say that they are necessarily more successful because it may turn out that many exciting new health findings are fads after all, and the wait-and-see attitude paid off. This is no excuse for management to constantly refuse new ideas linked to new health and nutrition-related theories. However, it is easy to comprehend that one should be careful with new nutritional and dietary recommendations: when looking up "list of diets" on the Internet, one can find reference to almost 100 different diets; some of them have similar or identical underlying nutritional rationale, others are different in their approach.

It is the professional duty of the nutrition expert in the R&D organization to distinguish and decipher these multiple approaches and seek out the differentiation that brings value to their company. All this clearly highlights the difficulty for a company, especially a food company, to have an organizational set up of its R&D group that is always up to date at any given moment. Sometimes such groups and their individual representatives appear to be outdated and chase after projects of the past like they were archaeologists or anthropologists, just trying to hang on to their themes because those are the only ones they really know. This is clearly not an easy situation and has no easy fix. The food industry is traditionally conservative, and this holds true for large corporations as well as smaller, almost family-sized companies. Consumers don't want, and certainly don't, expect outlandish novelties related to their food and react rather negatively to attempts by the food industry to bring out totally new food concepts. I should, however, add that it is probably difficult or close to impossible to discover, develop, and launch a completely new food product that has not been known, at least in its basic approach to mankind. That doesn't include the legal difficulties in obtaining approval from food authorities anywhere in the world.

3.1.2 Food is a conservative beast

You may argue that because the industry is so conservative and nothing really new has been brought to the marketplace since, well, who knows, it should be simple and straightforward to structure, set up, and organize any type of food industry R&D group. You may equally argue that because the area of food research is so conservative and novelty-adverse, there should not be much of a food R&D in the first place when it comes to effort and resources. You may be right on both accounts but only partly and I will discuss these apparent contradictions throughout the book.

So, let me get back to the typical setup of a food industry R&D group and attempt to carve out differences as well as similarities between smaller and larger companies and R&D groups. From many years of personal experience, I can say with authority that there are many more similarities than differences but let's get to this step-by-step. I have already mentioned that one of the first starting points when it comes to organizing a food R&D group is to structure it by ingredients and technologies (or processes, if you prefer). When I say ingredients, it is really a long list that not only comprises the major food ingredients, the "macronutrients" (i.e., lipids, carbohydrates, and proteins) but also the minor ingredients, the "micro nutrients" (i.e., flavors, aromas, vitamins, minerals, texturing agents and emulsifiers, stabilizers, and probably a few more such as colors and antioxidants). In a large food R&D you might find a group of experts for most of these ingredients, some smaller and others larger, depending on the weight and importance that the company attributes to each of these.

3.1.3 Small is beautiful, or is it not?

Smaller companies typically would concentrate the fewer resources they might have and group macronutrient experts together in one group working on the major food nutrients; they would also typically group texture and emulsifiers and other expertise such as emulsions, foams, dry foams, and similar know-how together in one group. Large companies would typically have expert groups for each of the listed ingredients, macro as well as micro, and often struggle when trying to not only coordinate these efforts but also bringing them together in any given food matrix. There is something to be said for working jointly and more holistically from the start, but I have seen and lived such individualism in large companies to the detriment of the hoped-for result. It requires good knowledge management and managers who have a vision for the overall end result. In larger corporations' R&D groups, such individual expert groups are joining forces by being grouped together in a department structure, which on the one hand makes perfect sense but on the other adds yet another hierarchical level to the most likely high number of levels that already exist in every company, small or large. I will discuss the topic of hierarchical levels later in this chapter in more detail.

So, food R&D organizations would typically have departments that deal with food science as well as food technology because the one cannot really be separate from the other. Working on a project that deals with lipids in the food matrix requires knowledge in both, ingredient as well as its incorporation in any type of food matrix, which is typically more a food technology–based know-how. The same goes for any other macro- or micronutrient and definitely is a strong reason for a joint approach. Beyond the need and logic of combining science and technology, there are at least two more needs that can only be dealt with in dedicated knowledge groups.

3.1.4 Ingredient is king

The first one is the impact that any type of ingredient or a combination of ingredients has on our body, metabolism, and health. Questions related to this topic are typically dealt with in departments that are specialized in nutritional aspects, grouping together nutritionists, experts in body metabolism, and other health-related topics, most often under the heading of a nutrition department. Other companies call it home economics or other more or less fancy names such as human health or food benefits. The name is not important as long as everyone inside and also outside the organization understands what it means. It is also important for everyone in the organization to understand that the one (food science and food technology) has no real reason to exist without the other (nutrition), which, by the way goes both ways: nutrition alone cannot sustainably exist in a food R&D organization.

I have seen time and again ugly disputes and almost battles between these groups when it came to highlight one's importance and supremacy in the organization. More recently, however,

I have experienced a vast improvement of the situation, probably based on factors such as better mutual understanding as well as strong impact from management to not fall back to the older habits of antagonism.

The second, extremely important impact coming from the sheer fact that the food R&D group works on ingredients is their quality- and safety-related aspects. The mantra is, and has to be, that whatever work is done in a food R&D organization, it has to adhere to strict and well-defined quality and safety standards. Well-defined means that these standards are typically set forth by government organizations or associations, such as US Department of Agriculture, US Food & Drug Administration, European Food Safety Authority (EFSA), or others. These organizations all have similar rules that they put out regarding the types of ingredients that may be used in which type of food and at which maximum amounts. All these recommendations or rules may differ in different parts of the world and thus make putting these into practice an often extremely cumbersome and even tricky task for those in the companies who have to adhere to these rules.

3.1.5 Quality and safety are not everything, they're the only thing!

That is the main, or probably only reason, why every food company has set up a department dealing with quality and safety (Q&S) topics. Typically, such a department is part of the R&D group. Q&S is, by the way, one of the rare departments, which has eye-to-eye exchanges and meetings with similar departments or expert groups in other, competing food companies. Such exchanges would typically happen at large industry association meetings and, to the best of my knowledge, never in one-on-one meetings. The latter would be critically viewed, rightly so, as breaking antitrust rules that have to be respected by all means. I can say that I have never seen nor participated in bilateral meetings with representatives from competition because the negative outcomes of this happening would outweigh the possible gaining of knowledge improving my company's product quality. I have to emphasize that I am reporting personal experience but can, of course, not exclude that such meetings *can* happen. This is not to say that gaining knowledge from competitors in other, legal ways is not a good thing. However, the emphasis here is on *legal*. I shall discuss this in more detail in Chapter 4.

Ingredients-related competence groups in a food R&D group are one side of the coin, the other one being competence groups linked to technology and process know-how. It is sometimes difficult to understand that these subjects are treated in separate groups from the ingredient groups and in smaller food companies this would typically be dealt within the same expert group. However, large food companies have existed for many years, some of the bigger ones even 100 to 150 years, and were mostly created based on a technological discovery by the original founder of the company that was a solution to an urgent consumer need. As an example, and this probably comes as no surprise, I would like to mention Henri Nestlé, the founder of the Nestlé company, who saw the nutritional needs of babies during the 1860s and who had the idea of a technical solution for the need. This technical solution was based on drying liquid milk-based mixes, which, in these days was no easy feat. Thus, the company he had created, was to become the first food company with drying at the core of its technology base. To this day, almost 150 years later, drying is still a core technology for the Nestlé company and many improvements, discoveries, efficiency enhancements, product improvements, sustainability increases, and a few more were important milestones during this time. It goes without saying that the company has introduced many more core technologies such as extraction, emulsification, pasteurization, sterilization, aseptic filling, and milling, just to name a few.

3.1.6 Technologies are always product related

Such technological expertise was always built around a product, new or improved, and the invention of Nescafé in the mid-1930s was a typical example for this. Two technologies, extraction and drying, came together and resulted in a new type of product: soluble coffee. There are many more examples to be found in the industry, such as the combination of ingredient knowledge (oils and fats, lipid-reactive enzymes) and mixing and emulsifying knowledge, as invented on an industrial scale by the Unilever company in the 1960s and thereafter. The Velveeta Company, later (1927) to be purchased by the Kraft Company, combined their knowledge on cheese making with emulsification know-how and related technologies to create products such as processed cheese under the name of Velveeta. You can find many industry examples, which always point to the same conclusion: in the food industry technologies were never invented in isolation and without having a specific product or product group in mind.

This is different in other industries and even more so in the case of university research. Even food-related university research is often detached from a specific product in mind and is more interested in what can be called *unit operations*. Such unit operations can be related to mixing, crystallizing, milling, high pressure, electrical pulses, magnetic pulses, light pulses and the like, in general; however, all are linked to heat or mass transfer in one way or another.

As just mentioned, other industries, such as electronics, automotive, movies, and entertainment, often conduct research that is not yet intimately linked to a possible product but is more of a future enabling nature. One can sometimes find pockets of such an approach in the food industry's R&D groups but they are typically not well perceived and received by the industry's management, especially top management. This is certainly linked to the fact that there may not be an immediate value in such research and its results and therefore there is neither enough understanding nor the necessary patience to wait out applicable results of such an approach.

3.1.7 What's my project worth?

This brings me to an important point in this discussion. In the preceding chapter, I briefly mentioned the topic of value of R&D work and which is in my eyes one of the most controversial topics in debates between management, again typically top management and the R&D group of any food company. I do not want to delve in more detail in the value discussion and suggest that as a representative of an R&D group you may want to consult my other book (Traitler 2015) and learn more about this so important topic because it is decisive that any member of the R&D group can and must justify his or her existence in this important company of yours. The only important message that I would mention here is to realize that when it comes to claim one's part in success and which contribution R&D had to the final outcome of a successful product launch and that had its origin in the very R&D group, the discussions can become tricky, to say the least.

3.1.8 *Cui bono?*

Consequently, I shall terminate this section about a typical food R&D setup by discussing and analyzing the question of who profits most of the typical R&D setup described? The politically correct answer is: the consumer and everyone around the consumer such as professional operators, distributors, and retailers. And to some degree this is probably true, but then the real benefactors, the "profiteurs" are really to be found in the company itself, and more likely than not at the business and management level. This may sound strange to you, but in my eyes, this statement reflects a lived experience, not only in my former company but in most food companies' R&D organizations that I had contact with. Let me explain. The CEO of the company needs and certainly wants to have as much control as possible over most or all the various branches of the company, including R&D. Because he (more seldom she) cannot be everywhere and be informed about every little detail, there is a need for "good soldiers" who act rather on his or her behalf and execute his or her wishes and, of course, his or her visions.

I am not suggesting that these people are simple yea-sayers, but there is some of this yes-man (yes-woman) attitude to be seen on this level. So, like all the other executive board members, the CTO who is in charge of R&D is one of them and wants (needs?) to please his or her boss so that some job longevity is almost guaranteed. You may argue that longevity depends on performance and, in part, you are right, but only in part. Performance is important; however, loyalty and executing the boss' orders appears to be even more important at that level. Maybe it's even important at lower levels and could go through the entire organization? I have no definitive answer to that question but there might be something to that. Up to you the reader to draw your own conclusions. The consequence of all this is that it often appears that making your boss happy is more important than having a successful project, even if the latter should be something that makes your boss happy. The other consequence of this is that the CEO, or CTO for that matter who puts "his (her) people in place" have more direct influence and control of what is going on at any level of the organization, clearly pointing to the answer to the question of who profits most of the setup? It's management and not always necessarily the consumer. This is something that needs to be changed; Chapter 10 will discuss this in much detail and show alternatives to the present day R&D setup in the food industry.

Figure 3.1 attempts to depict a typical, simplified R&D setup in a food company. Note that the detailed setup can have many more levels and variations and the figure does not show administration and human resources at all.

3.2 THE PEOPLE IN THE FOOD R&D

People are our most important assets! Often said, maybe even more often heard and, in my eyes, only partly true. I believe that the saying should go: "The right people are our most important assets." You may of course debate, whether people can really be assets but, for the sake of simplicity, let us assume that they are. In *Food Industry Design, Technology and Innovation* (Traitler, Coleman, and Hofmann 2014), I have discussed the topic of people in the food industry quite extensively and I will focus on people in R&D in this chapter and this section here.

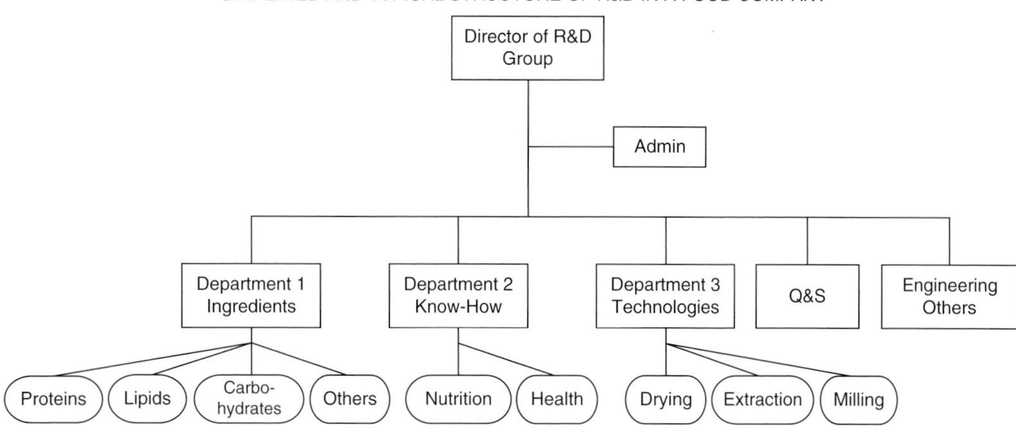

Figure 3.1 A simplified food company R&D structure.

3.2.1 Do I stay, or shall I move on?

It is to be expected that a certain type of industry and especially a certain type of corporate structure and culture attract a specific type of people. The food industry and its R&D organization are no exception to this. It is difficult and maybe also not really legitimate to characterize in general terms an entire group, the "population" that one can find in an R&D organization but let me try anyhow. And I do admit that this view has certainly its limitations. Based on my experience I see two major classes of people in the food R&D, and maybe you do find them in any organization in similar ways:

1. The "hoppers"
2. The solid "stayers"

The hoppers are those who cannot wait to get to their next job assignment, whereas the stayers form the solid backbone of longevity in the organization. I do not put any value classification on either, because both types are dearly needed in the organization, probably in any organization. Would there only be hoppers or only stayers, the organization would simply not function. Especially an R&D organization needs a strong backbone of stayers. However, if this gets out of balance, the organization runs the danger of not being open enough and sends its representatives to other parts of the company, be it within R&D if the company is big enough to have more than one R&D center or without in functions such as manufacturing or packaging in the area of production.

3.2.2 Twenty percent! Are you out of your mind?

This raises an important question: what's a good turnover of people in an R&D organization? By turnover I mean the newly hired, those who leave the company (on their own will or are being let go), and those who go to other parts of the company. I personally experienced a situation a few years ago of turnover of close to 20 percent, and this is per year! In other words, out of 100 people, 20 moved on to other functions in the business, mostly, however, training functions without guarantee of ever being taken back into a more permanent position. This apparently had gone on for a few years already and had created quite a stir, to say the least. Initially there is a lot of enthusiasm to be observed but this is quickly taken over by a realistic assessment of the overall situation and much frustration and disappointment. I was put in the position to run this particular R&D center and, after some assessment and developing a plan how to solve this clear mismatch, I totally stopped any transfer for the better part of one year.

Now, this bears the question, what is the "normal turnover"? There are no scientifically based numbers to be found and if there were, they are probably all based on situational observations and analysis of real-life situations. Based on my own experience and in agreement with many of my former colleagues, a number of 3 to 5 percent of turnover per year is probably a healthy one, with 5 percent being on the high end already. Again, this is true for the food industry and especially its R&D part and may be considered differently in other industries. This range may even appear high for SMEs, which cannot afford such a personnel exchange rate and depend on the group of stayers much more than a larger company, which has many more outlets into the business for its people.

It took me at least two years to come into a more appropriate cruising speed in terms of people turnover and had to fight a few more years to deal with the many people that were sent out with half-hearted promises during the prior years. We ultimately succeeded, together with the entire management team, and could thereby stop unhealthy attrition of our organization.

Figure 3.2 illustrates the various ranges of people turnover in a food R&D organization, based on personal observation and experience.

Let me get back to the types of people that we typically can find in a food R&D organization. Within the group of the hoppers, there are those who have strong belief that they are so great that everyone in the organization must have seen their talent after probably 6 months of them being in their present job and that after 12 months or so they not only deserve a promotion but actually should receive it and can move on to new grounds. I would call these, for lack of a better descriptor, the "super-hoppers"; typically these are people with a strong ego, probably full of themselves and with a super strong belief into their capabilities, possibly overestimating these, sometimes, hopelessly. Maybe not too many of them can be found in a food company R&D group but they do exist. I have seen a few.

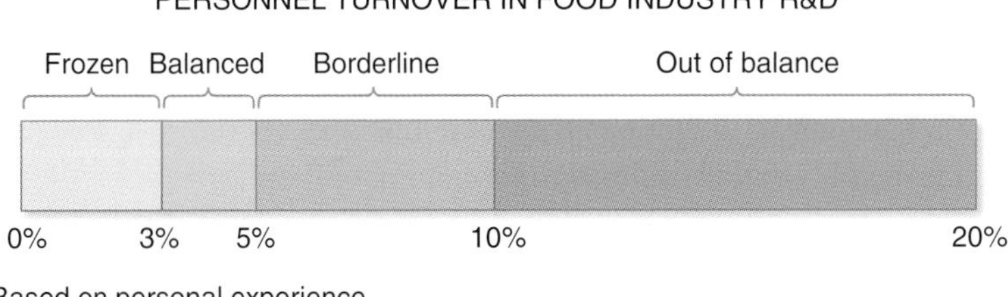

Figure 3.2 Turnover of personnel in food industry R&D.

3.2.3 More hoppers

Then you have the other type of hoppers, let me call them the "opportunity-hoppers." These are the flexible people in the organization who are always on the lookout for a possible new position while they typically loyally hold on to their current position and the projects that come with it. These form the majority of the group of hoppers and are, in my experience, a good group of people to work with.

And then there is a third group of hoppers, a group that I would call the "nepo-hoppers," where *nepo* stands for *nepotism*. Every company has these types of people; they are probably more numerous in positions outside R&D where profound university education and experience is maybe less of a requirement. I say this carefully because I do not want to belittle business functions outside R&D at all. Again, these people exist and they are probably the ones that annoy and disturb the system more than any of the others. But from experience, the best way to deal with them is to arrange yourself with the situation, or as I would say, "don't even ignore them" (meaning one level higher of "ignore them").

There may be other types of hoppers that you can possibly identify in your organization, but I believe that I have captured the three major subtypes.

3.2.4 More stayers

On to the group of stayers! Also here you can find different subtypes. The first one I would call the "habit-stayers," being characterized by strong loyalty to their present work environment and especially their colleagues around them. These are people who do a lot together with their colleagues from work, they go and have lunch with them, possibly every day and certainly, if ever possible at the same table or in the same restaurant area, in the case the R&D center runs a company restaurant. If not, they may go out together for lunch, they carpool together, and they often do weekend activities together. Let me be clear: there is nothing wrong with this; if it's not the one fact that might question the efficiency of people in this subtypes of stayers: repetitiveness of the daily actions, possibly resulting in a kind of laziness of mind. I have experienced people who were members of such cozy cocoons becoming thought-lazy and not open enough for obvious new things lying right in front of them. On the other hand, such cozy closeness may lead to almost conspiratorial behavior of its group members, which, in turn, may lead to surprising results in their R&D efforts. Members of these groups tend to take their work outside, home, or to social gatherings and thereby possibly stumble across surprising innovations. This is not an easy group to deal with and should be carefully managed to not destroy the positive by interfering with their close-knit circles too much.

The next subtype of the stayers could be called "enthu-stayers," where *enthu* stands for *enthusiasm*. Enthusiasm characterizes a typically large group of R&D people who are so excited and enthusiastic for their work and the projects they can contribute to, based on the important know-how that they have acquired through studies and practical experience. They have achieved a certain degree of notoriety in their science and technology field and are getting much satisfaction from internal but even more so external recognition of their contributions. I can add as a small personal note that through quite a few years while I was working for my former company

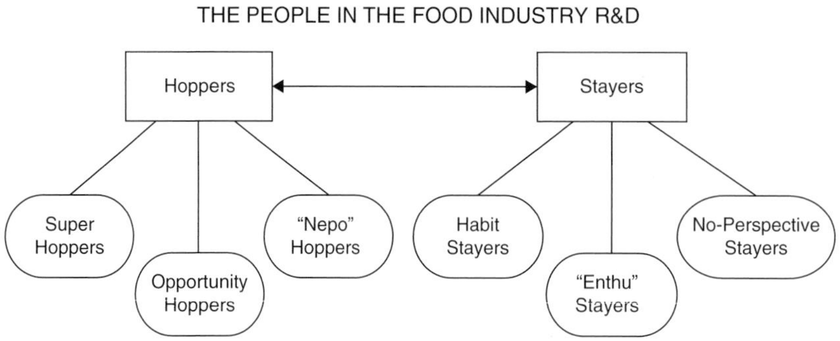

Figure 3.3 People in the food industry.

in a basic R&D environment I was such an enthu-stayer; it was great while it lasted, it was tough to get out of it; and it was even greater to discover that there was "land beyond the horizon." In general, this is a good group to have and most of its members are self-starters and need much less attention than the habit-stayers. This is not to say that one should forget them; they still need the recognition of their achievements more than anything else.

Lastly, there is a third subtype of stayers, which I would call "no-perspective-stayers." This is neither a good group to have nor to work with but such groups exist and therefore need to be dealt with in any R&D and other business environment. I am not necessarily talking about the non-performers who one can let go or possibly find a new position better fitting for them. Most often, members of this subtype develop from the habit-stayers, who remained too long in a cozy and nice position, not realizing that over time the opportunities for them became rarer and rarer. All of a sudden they may find out that there are no more new challenges there for them and they have lost all real work perspective. This is often a slow and hidden process that neither the person nor his or her boss or management have noticed when it was still possible to rectify and correct the situation. This is probably the most difficult group to deal with, and there is no one recipe for correcting and resolving the situation. From my experience, I believe the best way is to closely observe and, if necessary, mix up the cozy groups and confront them to new members and new and seemingly uncomfortable situations.

Figure 3.3 illustrates the various types of people typically found in the food industry and especially in its R&D organization.

3.2.5 Change can be frightening

This brings me to the next topic, or rather question: "Can people change?" Based on my personal experience, the answer is a straightforward "yes, they can change." The real question, however, is whether they *want* to change or whether they are even prepared to change. When someone works in one of these cozy groups and always has been happy, never had seen many opportunities for advancement in his or her job but has never had much criticism either, there does not appear to be a strong enough reason for the person to change. So, the answers to the questions probably are: yes, the person could change but does not see a need for change and therefore does not want to change.

On the other hand, everyone has dreams and desires what one could or should do, achieve, attain, and reach in the hopefully not-so-distant future. So, how come that you still find quite a large number of people remain in their present positions and not striving for advancement toward new and exciting positions? The answer is rather simple: it's fear of change, fear of coming out of the well-known comfort zone, and fear of finding new and still unknown territory. Fear of change is, in my experience the single-most important de-motivator for many people, especially for those who have the feeling of being well-established in their present positions. Why give this up? As most of us know, such fear is almost always totally unfounded because those who have come out of their comfort zone and have forcefully moved forward to new grounds are also almost always totally enchanted and satisfied with their new situations. But one only knows once he or she has crossed the bridge that seemed so menacing on this side.

In a typical R&D setup, the best way forward to change people's attitudes and get them out of their comfort zone lies in confronting them to exciting projects and involve them in works of discovery, invention, and innovation. If management can offer such challenges, everyone in any of the described groups and subtypes of people working in the company, and especially in an R&D environment, will want to move forward, highly motivated and all borders and classifications will disappear. It's not an easy task for management and for the company as a whole, but it's the only really successful approach to having a top-class work force, including scientists and engineers in the R&D group and keeping them happy, pushing forward to new discoveries and always carefully and prudently measuring the opportunities that lie ahead of them. This brings me directly to the next topic that I shall discuss and analyze in some depth in the following section.

3.3 THE ROLE OF DISCOVERY AND INNOVATION IN FOOD R&D

"Innovate or perish" is an often heard and widely used slogan in any industry, and this is also valid for the food industry. It can also safely be said that R&D is all about discovery and, of course, a few more things. Many books have been written about innovation, the role of innovation, and I have personally added my share to this list of publications and books (Traitler et al. 2014). I have extensively written about the role of design in the food industry and have described and discussed in much detail the interactions of design, technology, and especially innovation. In *The Food Industry Innovation School* (Traitler 2015), I have broadly discussed pathways toward a higher degree of success of innovation in an industrial environment and especially the aspect of driving innovation through complex organizations.

3.3.1 It's all about discovery

In this book and this section I shall emphasize the role and status of discovery and innovation in the food industry. Discovery does not necessarily mean innovation, although this is often the case, simply because much of the discovery in corporate research happens within the confines of company strategy and is not blue sky, like one could believe is the case at the university level. Let me look at the aspect of discovery first. *Merriam-Webster* defines "discovery" as the following:

- The act of finding or learning something for the first time
- The act of discovering something
- Something seen or learned for the first time
- Something discovered

The common denominators of the various definitions are: "find out or learn" and "for the first time." The find-out-and-learn part is the easy one, I mean easy to understand that that's what discovery is all about. The for-the-first-time part is more difficult to apply in the context of R&D. Although everyone would whole-heartedly support the idea that real scientific discovery is finding something for the first time, I would respectfully respond that you cannot say with absolute certainty that you personally found out for the first time. Yes, you go online and do literature search, patent search, or other searches and find out that nothing had apparently been found out about the idea that just came to your mind, your discovery. I am a big believer in the idea that there is no real zero-point for scientific discovery. I would dare to compare the situation of a zero-point of discovery with the big bang, although I have to admit that it's far-fetched. What I really want to express is the fact that every discovery by an individual or a team is always based on known facts, on past experience, on existing knowledge, and personal history.

What I would therefore add to the definition of discovery is the aspect of chance, surprise, and serendipity. For me, every discovery is almost always first of all a serendipitous event and only then comes the aspect of for the first time and the latter only in a narrow definition of having built on existing knowledge. And I do insist on the building aspect and defining it as "based on" would be far too narrow for me. So, in all, discovery is really about learning or finding something out, with a bit of luck and having built on and put together past personal or group knowledge, and the aspect of for the first time comes last, which may surprise some of you. This is not a philosophical discussion, but this is all about clearing the ground for the innovation part.

3.3.2 It's all about innovation, or is it renovation?

Not every discovery leads to innovation, but all innovation is based on discovery. This is an important statement because it shows the simple correlation between discovery (finding, by chance, building, for the first time) and innovation (something surprising and useful for the end user, found by chance, having built on past or present knowledge, probably for the first time). I jumped a bit ahead of myself here by offering a kind of definition, as I see it, for *innovation* by listing these various elements. You may add as many as you like and see fit, however, I strongly believe that the important ones I have just mentioned. However, the food industry R&D work is not only built on discovery and innovation but to a large extent on what is usually called *renovation*. Renovation is a form of innovation and mostly consists of moderate and careful modifications of existing products and processes. Innovation rocks, while renovation does not rock the boat but just slightly shakes it.

I can safely say that renovation is the food industry's preferred one of the siblings called innovation and renovation. Why is that so? Well, to begin with, it typically happens faster and costs less. There is less money at stake in case it flunks and can be written off more easily and quicker than a more costly and more lengthily developed innovation. Renovation is by nature less radical than innovation and can be seen as a form of incremental innovation, therefore going only baby steps at any given time. In other words, innovation is often perceived as more of a revolution, while renovation is considered to be evolutionary. The main argument, however, is the argument about market and consumer acceptance: is the marketplace, is the consumer, ready to accept fairly radical innovations in the area of food or is it better to lead all actors in rather small baby steps toward gradual product improvement and thereby not risk to antagonize the end user?

3.3.3 Size matters

In actual fact, management in the food industry will almost always encourage and support small-step, incremental innovation, or simply put, renovation. So, the slogan that I used should go: "renovate or perish." Maybe this sounds a bit disappointing to some of the readers, but it does reflect a reality in the R&D groups of the food industry. I should, however, draw a slightly more differentiated picture with regard to the desire or need of pursuing innovation or renovation, depending on the size of the company. I have personally experienced a large degree of differentiation especially in the area of packaging but also product and process development. Typically there is a strong inverse correlation between the size of a company and its balance of portfolio of innovation versus renovation. Or in other words: the smaller the company, the more innovation and innovative products one can find in that company, while bigger companies tend to favor renovation and renovation-based products and processes.

This phenomenon can partly be explained, especially in the world of packaging, by the simple fact that large companies typically produce in large quantities on large equipment such as printers (11 print-head oversize machines), King-Kong–size carton erectors, flow-wrappers, or anything else along these lines and therefore have quite naturally little room for maneuver, let alone try out new things. They are the converter companies that produce for large end-user companies stock materials. Smaller companies seem to have the privilege to be more flexible, produce on smaller machinery, and can offer custom, bespoke materials to the food industry, which makes factory trials easier and not only more manageable but also more affordable. This size-related trend can also be observed in most food companies producing products for the end users, the trade, retailers, and ultimately for the consumer. Unfortunately this argument of too bulky size to try out something new on a larger scale is often used by the industry to cut more radical innovation short and concentrate on safe and simple renovation.

Figure 3.4 depicts this relationship between company size and degree of innovation versus renovation in a simplified and illustrative way.

3.3.4 Here's a way out

Some large companies make up for this by having created product and process development and especially implementation groups, sometimes called *application groups*, in the proximity

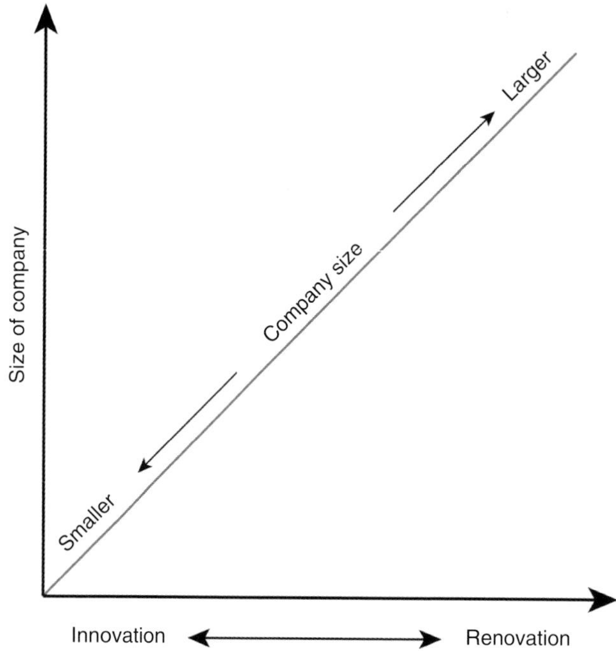

Figure 3.4 Innovation versus renovation as function of the company size.

of the production plants, as a kind of missing link between R&D and manufacturing. Such groups have as their most important task the pursuit of implementation and industrialization of projects, be they innovative or of renovation character that result from R&D. If trust levels are high and there is enough mutual respect between the R&D group and a corresponding (meaning product-wise) application group, such collaboration works pretty seamlessly and without major problems or hiccups. However, if it doesn't work well, it can actually become counterproductive, which leads to inefficiencies by duplication or right out hindering of advancing. Luckily, such situations are rather seldom and require management to play a constructive and unifying role. I have personally experienced a great number of great project results that came through the appropriate alignment and complementarity of R&D and the application group. Apart from the fact that it's probably a setup that only larger food companies can afford, I have to say that it's a really neat and useful setup to have and to work with.

3.3.5 What would the consumer say?

There is still the one important question that remains to be answered: "Is the consumer prepared for innovation that results in new food products?" By new food products I mean those, which are radically new, like it used to be the case for soluble coffee in 1938 or margarine through fat hardening and later through enzymatic interesterification of lipids in the mid-twentieth century or energy drinks in the 1990s or similar types of really new food and beverage products.

More recently so called "superfoods," although they always existed, came into the marketplace and are more and more to be found in the health and nutrition corners of supermarkets and specialized shops. Superfoods are food products that contain high amounts of active ingredients such as vitamins, minerals, phytosterols, antioxidants, and such and have, what I would call "hearsay" efficiencies, although some or many of them are based on sound nutritional science.

The simple answer to my question is: yes, consumers are prepared to test something rather radically new but not all of them. It's like with every new product of any type in the marketplace: you find the early adopters, the mainstream users, and the late followers and probably a few more subgroups. Against common belief, food products, which result from radical innovation are not necessarily refused by the consumers. On the contrary, more and more people can be found who buy such products. Times are gone when food products followed just one or two principles such as tasting good and being healthy. Today you find consumers for just about every trend, fad, fashion, and latest hype. I am not saying that this is good, but it's a reality and the industry and its R&D group follow these trends, although most of the companies do this carefully.

Chapter 6 will extensively analyze and discuss questions and issues related to consumers and the consumer perspective for potential change that is being proposed in Chapter 10 and this book. Neither of these chapters, and especially the assumptions, ideas, and proposals that will be discussed pretend to "speak on behalf of the consumers" but will attempt to take different possible consumer reactions and trends into account. The reader can make his or her own assumptions based on their own personal and professional background and experience. I am far from pretending to have the prevailing opinion, but only speak from my own experience and many years of observation in the industry.

3.4 ADDITIONAL PERSONAL OBSERVATIONS AND R&D-RELATED STORIES

Building on what I have just expressed, let me recount stories and experiences that are related to the R&D world in the food industry. Some of these stories you can also find in my other books and I shall complement them with more stories and observations, which are directly R&D related.

Most of the stories and observations, if not all of them, are project related and are complementary to what was analyzed and discussed in the previous chapter.

I will relate the stories in a slightly systematic manner by grouping the various types of projects that are typically to be found in the R&D group of a food company. Actually, I do believe that this classification is also valid for the R&D world of other industries, but I shall focus on the food industry here. By looking into project-related stories, which are based on my personal experience, I could make out eight different types of projects. Here we go:

- The briefed, strictly business-based and justified projects or the business project
- The secret project
- The pet project

- The never-ending project
- The trial-and-error project
- The please-someone project
- The defensive project
- The knowledge-building project

I am realistic enough to admit that there may be even more types of projects and the reader may add as many as he or she sees fit and that I have unknowingly or conveniently forgotten. The stories that I have chosen really serve to give some personal insight into these different types of projects and render them a bit more human and tangible.

3.4.1 The business project

As I have discussed, all projects in any company in the world of R&D or outside should always be based on two major reasons: strategy and business need. In theory, this is exactly what a briefed project does; it follows strategy and business need. Or so you may think. Because strategy is defined by people and because business need is equally defined by people, it's ultimately these people who decide whether or not a project is going to be defined and carried out. Moreover, it may be the same people who decide, whether a project has come to a sufficiently successful termination, good enough to be going to the next level and ultimately being submitted to the consumers in the form of a product or service. I discuss this topic of driving innovation through complex organizations in my book *Food Industry Innovation School* (Traitler 2015). Suffice it to say that because there are people involved all along, all smart ways forward to achieve success have to be people related ways and therefore can often be of irrational rather than rational nature.

My story dates back a few years and recounts a personal experience. My former company, not totally unsurprising, was—and still is to this day— active in research and development related to coffee, especially soluble coffee. I had the pleasure and certainly also honor to be put in charge of a big project, and by big I mean really big, with a large number of R&D people (full-time equivalents as they are also called by human resources), spread over at least five countries and closely followed and highly scrutinized by management, including top management of the company. This was not surprising given the nature of the project (coffee) and given the number of the resources put behind the project as well as its internationality. Oh yes, the topic was improving the perception of a good coffee smell above the cup of a reconstituted coffee. I am not giving any strategic or other secrets away here, just to illustrate what we tried to achieve and how we tried to achieve this with how many resources.

We were made to believe that we would have approximately 24 months' time to achieve the goal of a perceptible whiff of coffee smell above a nice cup of soluble coffee. You may find this awfully long, but we (the entire team) felt then that this was probably far too short, especially given something that I have not mentioned yet; this was not the first time that such a project was performed and such a result was requested by strategy, business need and, last but not least, top management. Anyhow, the project began with much enthusiasm and hope and with many great ideas that were tested and put into practice. At the end of the day, any type of food-related project, and especially this one can only be evaluated by tasting sessions, or as in this case, sniffing sessions. The two big shots who were sponsoring the project and who were also members of the company's top management were heavy smokers (cigars that is) and whenever there was a project review followed up by a sniffing session, these two managers brought along with them not a whiff of coffee smell but one of heavy cigar smell. You can imagine, how this influenced the results and how they could not detect any noticeable coffee aroma in such sessions and the project was put to rest after 1½ years or so.

This might sound like I am looking for excuses that we did not succeed in such an important endeavor, and maybe I do; on the other hand it demonstrates that even for super important projects that bind a lot of resources and clearly follow business strategy and needs, professionalism and rigor are often put aside and personal habits and managers letting their personality and personal habits win over good professional behavior can be observed in such circumstances.

The learning is simple: because a project is solidly briefed and supported by everyone in the company does not mean that it is also professionally followed and judged.

3.4.2 The secret project

A secret project can become a nightmare in case you are caught in it. Luckily I was never personally involved in such a misadventure, but I have seen some of my former colleagues suffer quite a bit in such a situation. A few years ago a new managing director of an entire market (in some companies this might be simply called country or region) wanted to completely redesign the packaging of an entire line of products. Normally, this would have meant to involve quite a few people inside as well as outside the company such as the person responsible for the business in question, the technical person responsible, the packaging experts, the manufacturing person responsible, marketing, sales, in other words, the entire organization plus resources from outside such as market research/consumer research and not least the trade.

None of this had happened in that instance; the managing director decided to go on his own and only involve a designer who was not privy to designing packaging. The project was kept so secret that none of the people I have just mentioned was really aware of anything but somehow, this project would have to be executed, should the stage of paper exercise ever be left. Here comes the solution: the person who ran the factory small-scale developments and trials (i.e., the head of the local application group was put in charge) and he had to make contact with packaging converters, all in secrecy, had to prepare the ground in the factory to eventually be ready for future manufacturing of the new packaging. The problem was that he was led to believe that he would have to leave the company in case he said anything before the official announcement of the project by the managing director. That's a real tough situation and that's what I meant by saying that such secret projects easily may become nightmares. The outcome, by the way, was a disaster and the managing director eventually left his position.

The real learning here is of multiple nature: first, try to avoid to get caught in such projects; secondly don't give in to blackmail (if you can, it's not always easy, sometimes outright impossible); and, third, if you are member of such a project, better make sure that it's a successful one!

3.4.3 The pet project

Pet projects are mostly coming from the top. Henry Ford's "Edsel" is really the best historic example for such a type of project, but the food industry is full of them too. Pet projects are really top down and mostly come from the CEO or the direct layer below the CEO. My story takes me back quite a few years, so I can rather openly tell it because I do not divulge any secret here, rather a curious and human story after all.

One of the CEOs of my former company had a dream, at least this one was rather public and known to all: he wanted R&D develop a savory (i.e., salty snack bar). Somehow, this was a really emotionally driven idea, probably reminding him of his younger days when he was an apprentice and worked in the culinary area. The snack bar should taste like lovage (*Levisticum officinale*), in German also called "Liebstöckel" or "Maggikraut." The latter name reminds of the taste being similar to the typical Maggi Seasoning®, an old product, well-known especially in Germany.

So, all this came together; the CEO being German, the past proximity to culinary products, especially the range of Maggi® products, and a strong belief that the world of consumers has certainly waited for a savory snack. Bingo, here you have all the ingredients for a "pet project." Needless to say that he "bothered" every new research director with a renewed request to put resources behind the development of such a product, and I do recall that I was involved in such development cycles at least once, maybe even twice. My memory fails me here, simply because it has become almost a running gag: oh, he wants us to develop a "Maggi-bar," again? On the other hand, you can hardly refuse the wish of the big boss and there are actually always people who feel honored and flattered to work on a project for the top shot, even if it's been asked many times over already. On the contrary, it gives the individual the feeling and strong belief that all prior developments were flawed and now we will show them and especially the CEO.

Needless to say that the world did not wait for a savory, salty, seasoning-like tasting snack bar and, once the CEO had retired, so did the idea of his pet-project.

The major learning is again twofold: first, pet projects have a shelf life pretty much identical with the one of the requester, and secondly, try to stay out as much as you possibly can without endangering your standing and your career in the company.

3.4.4 The never-ending project

The existence of the so-called never-ending projects is maybe the biggest surprise for me when it comes to the world of projects and how they are justified, supported, and carried out. They do exist, they once were begun, and then ramification after ramification was added to the project, making traceability virtually impossible. So, such a project gets a life of its own and resources are put behind such projects as the new findings that occur during the progression of such a project always lead to the agreement by management to give it yet another try, great things are to be expected just around the corner. Reminds me of the time when I was a little boy of 6 years or so. My parents liked to hike in the mountains and for a 6-year-old this can be pretty boring. They knew that I loved to drink water, I did everything for water, that much I loved it. So they invented the "water-carrot" stick for me. Whenever I stalled in a hike up a mountain, they told me that right after the next corner there would be a water fountain, and I would be able to quench my first and drink my so much loved water. They were either lucky or knew the terrain pretty well, because most of the time there really was a fountain and I went on and on until we reached the summit. Here's where the similarity between the two situations, the never-ending project and my water-carrot stick ends: my ascent (progression) came to a halt when we had reached the summit, never-ending projects never seem to reach any kind of summit but get lost in the cloudy heights of summits never seen.

The oldest never-ending project I ever have come across was a project based on extracting ingredients with especially antioxidant properties from a plant source. When digging into company archives in the early 2000s, I came across a letter dating back to the summer of 1955 (!), requesting to start this project. I was really flabbergasted by this, because I was personally involved in this project a few years prior to my archive search, and I know for a fact that there are still people in my former company who appear to have a keen interest in the projects and pretend that there is still something to be found that was not discovered yet. I will let you make your own conclusions.

All I suggest as major learning in the context of the never-ending types of projects: because they exist (in all companies, in every industry, maybe not as old as my example) make best use and participate in such a project for a short period of time and learn; learn about the content but learn to make new connections and use the vast network that inevitably has been erected around such a project.

3.4.5 The trial-and-error project

This is probably one of the better and useful types of projects in any R&D and is most likely underused and underrated. Such projects are ideal to try many things out, quick and clean, requiring only a few resources and obtaining preliminary results in the shortest possible laps of time. There is no harm done having many of such projects, and every R&D group, especially in the food industry should actually run plenty of these. Typically, the fear of everyone involved is that once an intermediary and preliminary results is achieved it might be over and, if not pursued, one becomes part of a "loser team." The latter, especially when accumulated over time hurts the most, and for especially that reason, many people in R&D don't like to participate in such trial-and-error projects. And yet, this is probably the most efficient use of resources in R&D and should always be seen complementary to the longer-lasting, more profound, and meticulously briefed projects. The world of culinary products works a lot with this type of project in their development and test kitchens, and makes good use of all resources, experts as well as equipment.

The major learning here is simple: push for such projects and actively participate in them.

3.4.6 The please-someone project

One could confuse this type of project with the pet project, however, there is a subtle difference: while pet projects typically are initiated by top management, especially CEO or second level, this type of project, namely "please-someone" can come from every superior level in the organization, sometimes even from outside, such as from the boss's wife or husband. The other difference can be that often the please-someone project is initiated in, what I would call, "anticipatory obedience" mode. The employee believes that he or she has heard an idea expressed by the boss in a meeting, as anecdotic and far-fetched as this may be, and jumps on this idea and creates a project, only to be able to show his or her boss the wonderful results of this even greater idea.

Again, while the pet project comes from "above," and one often has no other choice than to obey and deliver, the please-someone project is entirely self-inflicted and therefore totally avoidable, unless the idea was really a great one and it was surprising that no one has thought of this before. This, however, is rarely the case but, because especially an outside person has potentially a different view of things, it could happen.

The major learning is a simple one: evaluate the idea carefully and in almost all cases, stay out of it.

3.4.7 The defensive project

Every organization and R&D group has defensive projects and they are almost a must have. From my observation, I can distinguish between two subclasses here: the defensive project that responds to emergencies in the market such as wrong dosage of minor ingredients in sensitive products, and the other type aims to generate patentable results, which can be used to strengthen the critical core of the company's patent portfolio. Chapter 4 will analyze and discuss this topic as well as a multitude of other IT-related topics. Both of these classes are extremely important. While the emergency-related class is undesirable, although it happens, and cannot be planned, the second patent protection class is part of the company's strategy and needs to be fully supported, not only by management but also by the individual employee.

The learning is simple: such projects do have to exist, resources need to be put into them, and for the individual this can be a tremendous learning, especially to work under time constraints in emergency situations.

3.4.8 The knowledge-building project

This type of project is the last one on my list, but I do concede that there may be more types of projects out there, which I do not have on my list. There is probably not much to be said about such projects, especially when they are carried out in a company's more research-focused R&D group. It's part of the continuous improvement drive of the individual employee, not only to keeping up to date his or her expert level but moreover, discovering, learning, and potentially applying new science and technology into all the other projects that are run in the company R&D group. On the one hand, this may sound like a routine activity similar to getting up in the morning, brushing teeth, and so on, and yes, it has some of this, too but there is much more and this is all linked to curiosity, discovery, and the drive and desire to learn and potentially apply new findings. One could consider such projects as the bread and butter of R&D and, from many years of personal experience, I can say that it is ever so exciting to be a part of this drive toward discovery and learning.

The learning is simple: if you ever have a chance to be part of it, grasp the tremendous opportunity!

Figure 3.5 illustrates the various types of projects and their main characteristics.

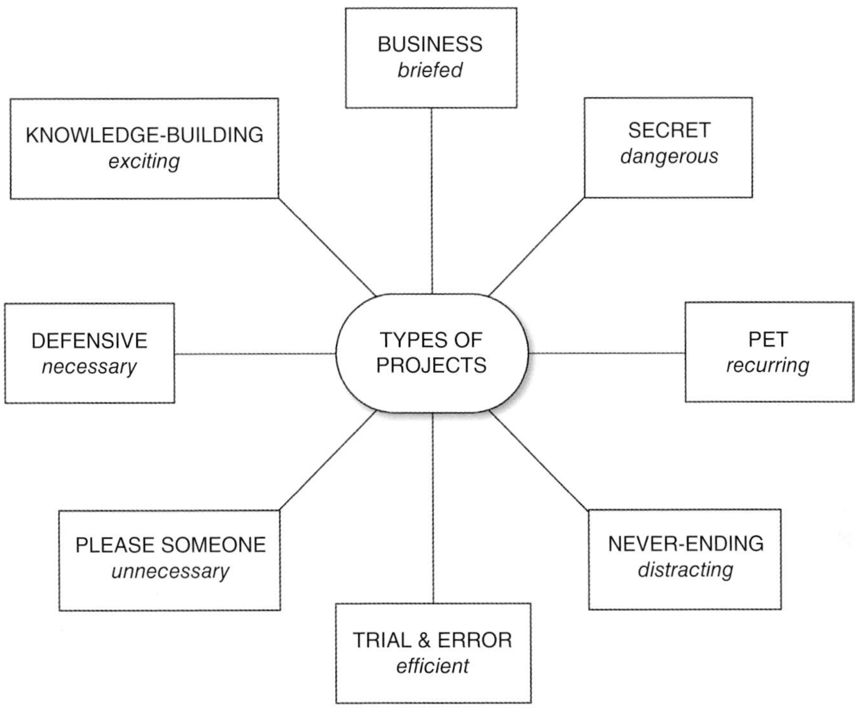

Figure 3.5 Types of projects.

3.4.9 Change is needed!

Let me terminate this section with one important observation that derives from the status description that I painted all along through this entire chapter: I personally not only believe but also am convinced that the present setup of the R&D group in the food industry, and I dare say every food company is suboptimal, and rather inefficient. The underlying idea of this entire book is to analyze and discuss this inefficiency in quite some detail and propose new possible approaches to an efficient, sustainable, affordable, and successful R&D organization in every food company, be it small or large or somewhere in the middle. In Chapter 5 I shall start this critical analysis and discussion on the need for a new approach to R&D in the food industry, and in Chapter 10 I shall propose and critically discuss possible new structures, contents, types of people, and expected subsequent outcomes of such a new approach.

3.5 SUMMARY AND MAJOR LEARNING

- A typical setup of a food company R&D organization was described and discussed. Historically and still to this day, the typical R&D structure is along the lines of ingredients, technologies, and general competences such as nutrition or health.

- The food industry is traditionally a conservative industry, therefore their R&D acts rather traditionally as well and can be described as having a conservative and rather careful approach to R&D of new products and processes.
- It was suggested and analyzed based on personal observation that the smaller the company and its R&D group the bigger the drive toward innovative solution, while larger groups typically focus more on renovation of their product and process portfolio.
- The role and importance of quality and safety (Q&S) was discussed because typically every R&D group in the food industry occupies experts in this field. "Q&S is not everything, it's the only thing!" should be and to a large degree is the main mantra in the food industry.
- It was mentioned that all technologies in the food industry are always product-related as the amount of investment that every food company makes into R&D is fairly small compared to other industries and limited resources have to be used pragmatically and in focused ways.
- I analyzed and discussed the question as to who profits from the present typical setup of the R&D group, and I suggested that it mainly helps management, especially top management, because they can exert better and more direct control.
- I introduced a discussion on "people in the food industry R&D groups" by suggesting that there needs to be a certain amount of personnel turnover, which should neither be too low nor too high. I proposed that the ideal rate of turnover of the total R&D population should be between 3 and 5 percent per year. Below such turnover, the organization runs the danger of becoming asphyxiated and complacent, while higher and ever much higher numbers, which I have personally experienced, take the organization out of balance.
- I analyzed and discussed the people in the food industry and suggested two main types of characters, split into six classes. The two main classes are hoppers and stayers. These may be subdivided into super hoppers, opportunity hoppers, nepo hoppers, habit stayers, enthu stayers, and last but not least the no-perspective stayers. They seem to be forming an almost natural habitat in every R&D organization and need to be managed carefully.
- Managing such people requires the possibility to propose change to those who maybe need it most and who the organization needs most to evolve, personally and especially professionally. Getting people out of their comfort zones is one of the most important tools in this endeavor.
- The role of discovery and innovation in the food industry was discussed, and it was discussed, this time in more detail, that size of an organization almost has a negative impact on the desire and need to discover and innovate; renovation is the preferred playfield of larger organizations, while innovation, most often out of almost survival need is the preferred direction for smaller companies to move forward.
- Ways out of this seemingly negatively impacting dilemma is the creation of smaller, practically acting units even within larger organization, which can concentrate on innovation more than larger groups would.
- I finally discussed a number of observations based on personal experience and told a few stories, linked, as basically everything in the food industry, to projects. I suggested eight

different types of projects: the *briefed* business project, the *dangerous* secret project, the *recurring* pet project, the *distracting* never-ending project, the *efficient* trial & error project, the *unnecessary* please-someone project, the *necessary* defensive project and, last but not least, the *exciting* knowledge-building project.
- Each of these types of projects can be found in the R&D groups of the food industry and all probably have their right to exist in one way or another, even if I have characterized some of them dangerous, distracting, or unnecessary.
- I terminated this chapter by suggesting that based on many years of personal experience and many detailed observations, the present typical setup of R&D in the food industry is, to say it nicely, suboptimal and to be more direct, outright inefficient. Change is required, and Chapters 5 and 10 will discuss such change in much detail.

REFERENCES

Merriam-Webster nd, "Discovery," retrieved from http://www.merriam-webster.com/dictionary/discovery

Traitler, H 2015, *Food industry innovation school: How to drive innovation through complex organizations*, Wiley-Blackwell, Hoboken, NJ.

Traitler, H, Coleman, B, Hofmann, K 2014, *Food industry design, technology and innovation*, Wiley-Blackwell, Hoboken, NJ.

4 Understanding intellectual property and how it is handled in a typical food R&D environment

Intellectual property is an important legal and cultural issue. Society as a whole has complex issues to face here: private ownership vs. open source.

Tim Berners-Lee

4.1 QUEST FOR INTELLECTUAL PROPERTY: AN IMPORTANT DRIVER

The traditional belief in any food R&D organization was, and to some degree still is, intellectual property (IP). IP stands for patents and patents need to be owned by the company, period. Having said this, the second part of this section's title should actually be, "the most important driver." When suggesting that the quest for IP in general is the most important driver for a company, and especially its R&D group, I would assume that most of the readers would agree to that. However, if obtaining IP of whatever kind, and especially patents, is pursued at all costs, stubbornly and in time-consuming manners, then we have to differentiate between what makes sense and what clearly goes beyond this overall goal of "quest for IP."

Let me, however, first describe briefly what in my experience falls under IP in the food industry and I ask for forgiveness, should I have left something out from the list. I encourage the reader to complete the list and discuss the individual items in more detail.

So, here we go.

4.1.1 Patents

Patents, especially process patents (i.e., specific technologies required to obtain products in new, surprising, and hitherto unknown ways). Like for every patent, there has to be a degree of novelty involved and the new to be patented process is such that a professional expert would not have easily discovered it. The difficulty to prove this lies in the simple fact that once you

Food Industry R&D: A New Approach, First Edition. Helmut Traitler, Birgit Coleman and Adam Burbidge.

see a new process laid out in a patent description, the professional expert could say, clearly in hindsight: "oh yes, that's obvious, I could have thought of that too." The latter is, however, irrelevant because it is hindsight. The two keywords are: novel and unexpected.

There are more types of patents, such as product patents and design patents. However, both of these types of patents are fairly weak to defend and the industry therefore strives for a combination of either of these with novel process, and preferably applies for product-and process or design-and-process patents or other possible permutations of these. As stated in the introduction, patents are only one, yet important, building block in building a strong intellectual property fortress in the company.

Let me briefly comment on the possibility of the "preliminary invention disclosure" of an idea. This is a simple way of obtaining a priority date for a patent that may (or may not) follow at a later date, yet reducing the overall validity duration of a final patent. Typically, such preliminary invention disclosure is valid for a period of time and consists in depositing a short, concise, yet sufficiently clear description of the basics of the invention and, for a fee, for instance $100 deposit, the document dated, signed, countersigned in the lawyer's office. If no action toward a patent application ensues during the deposition period, the idea can still be pursued, but the first priority date will not be valid any longer. This is a simple and underused possibility to get at least some protection of one's idea already at an early state without much effort and even less costs.

A more extended form of this invention disclosure is the "provisional application" or provisional patent application. This option is valid for one year but could be written extensively as if it was the almost-final patent application. The inventor has a lot more data and examples at his or her disposal and has a year to confirm certain claims that were not proven yet. The same applies here; if within one year no action is taken toward a full-fledged patent application, the priority date will become obsolete.

4.1.2 Recipes

A second extremely important and often overlooked IP element for food companies is recipes. Such recipes could also be defined by the term *manufacturing secrets*, provided that the recipe goes beyond the list of ingredients and covers manufacturing and processing as well. Every food company, small or large, sits on a wealth and multitude of recipes of all sorts in all food areas they work on. Some of such recipes have almost mythical status such as the basic recipe of the Coca Cola® beverage concentrate. Its origin is shrouded in the mist of history and is allegedly only known to the inner circle of management of the company, if at all. The principal value of the Coca Cola® Company is the value of its brand, which in turn is based on the basic recipe. There are similar examples in the industry such as the KitKat® candy bar, which is based on two complementary recipe elements: the formulation of wafer, praline (the cream between the wafer layers) and the surrounding chocolate, and secondly the geometry of the KitKat® candy bar. This product is a typical example of a smart combination of process-and-product IP, here not so much in the format of a patent or patents but more so through its long history and traditional recipe. This leads me straight to the next element of IP, namely trademarks.

4.1.3 Trademarks

Trademarks are principally based on notoriety, or rather having achieved the status of notoriety for a given product. Notoriety is a largely a matter of judgment but simplified: if 7 out of 10 consumers would recognize the projected shadow of a particular project shape or logo, one can argue that this product has a recognizable design and has reached a state of notoriety, which allows for a trademark. There are a series of further legal definitions behind the terminology of trademark, but for our discussion here, the important aspect is "recognizable design." A trademark gives legal ownership of the brand name and related product for a period of 7 to 20 years (i.e., up to a length of time equal to patent protection). Typically, trademarks are a lot easier to obtain and to support than patents, and moreover cost a lot less.

4.1.4 Trade secrets and secrecy agreements

Next on my list of elements of IP in the food industry is the area of trade secrets and secrecy agreements. To some degree, it could be argued that recipes constitute an element of trade secrets, however, given that recipes are at the core of every food company, I treated them separately.

Trade secrets are defined as follows:

> *A trade secret is a formula, practice, process, design, instrument, pattern, commercial method, or compilation of information which is not generally known or reasonably ascertainable by others, and by which a business can obtain an economic advantage over competitors or customers. In some international jurisdictions, such secrets are referred to as "confidential information," but are generally not referred to as "classified information" in the United States, since that refers to government secrets protected by a different set of laws and practices ("Trade secrets").*

Additionally,

> *The precise language by which a trade secret is defined varies by jurisdiction (as do the particular types of information that are subject to trade secret protection). However, there are three factors that, although subject to differing interpretations, are common to all such definitions: a trade secret is information that:*
>
> - *Is not generally known to the public;*
> - *Confers some sort of economic benefit on its holder (where this benefit must derive specifically from its not being publicly known, not just from the value of the information itself);*
> - *Is the subject of reasonable efforts to maintain its secrecy.*

A simple example of a trade secret is a disclosure of knowledge to a third party, typically an external partner for the intended purpose of establishing a future collaboration. Such trade secrets are then covered for secrecy by nondisclosure agreements (NDAs) or confidentiality agreements (CAs). Both, NDA and CA pretty much describe the same thing. There might be a slight legal difference, which has escaped me, but it has apparently never had any special adverse effect to the outcomes. Both, NDA as well as CA can be mutual or unilateral, the latter variant only covering the download of secrets from one party to the other. More often than not

I had signed mutual agreements, simply because both parties typically want to be covered for all eventualities.

There are a few important points to be considered in the context of NDAs or CAs.

1. Any such agreement contains a *time element*, which governs the duration of the validity of the very agreement. A rule of thumb is that 3 years' validity is probably the optimum time frame to keep exchanged information confidential. There are, of course exceptions to this in both directions, shorter or longer, and I have seen proposal for agreements that suggested time frames from 2 to 5 years, exceptionally even 10 years. The longer time frames might be imperative for collaboration areas in which for instance clinical studies are part of the joint project. However, for typical food projects, 3 years should be a good rule-of-thumb based duration.
2. Any agreement requires a *definition of the topic or topics*, which are going to be discussed under such an agreement. I have observed a tendency for these to be general from fear that by defining the topic to specifically one discloses too much even before both parties have signed. I have never encountered any problems in properly defining the topic of discussion and do suggest to be as specific as needed and as vague as possible. In case of mutual agreements this is to be expected from both parties.
3. The third point, sometimes a topic of contention, is the *geographic location of the court* that decides over the agreement in case it was breached by either party. Large corporations are represented around the world, so it is typically the larger, more widely present company that can more easily accept the court of law in the country in which the other party has its main business. In many years of having worked with third parties on projects, which were based on NDAs or CAs, I have experienced few tricky litigations. They were not tricky or critical because of the location of the court but because of them being disputed.

There are more elements in NDAs or CAs but they are of more legal nature, and most companies have preformulated agreements that sound similar to each other and have been checked and rechecked by lawyers many times over, so no real need to discuss them here in more detail.

4.1.5 Experts: Actions and results

There is more that falls under the general term of IP in companies, especially food companies, and that is the entire group of their own experts and their actions and results, inside and outside the company. From experience, this group is often overlooked and not really considered being a key element of intellectual property. You could argue that people are nobody's property (except for slaves); however, these experts possess the know-how that creates the IP of every company. You could also argue that this is obvious, almost trivial, and yet, it is never explicitly mentioned under the definitions of IP. I mention it here and emphasize it because it is always taken as granted and is, based on my own personal experience, almost always totally overlooked. So, the equation is simple: people (P_e) + know-how (KH) = intellectual property (IP). I used the term people's *actions*, and what I mean by that is simply their know-how, effort, and work that, inside the company go into every single project of any company and that leads to results that can be translated into successful products, processes, or services.

Outside the company, these experts are representing the image of their company, they are the proverbial "visitation card" that shows others what the company stands for and how its people represent the image. Yes, it's not only the people representing a company's image, but it's also a whole series of other items, first of all its products, its advertising, and its stand in the local, regional, and global community, the latter for the larger and largest food companies. However, the world begins next door and the image that is reflected by the individual expert, the carrier of the company's know-how is a critical beginning in the endeavor of projecting the best possible image of the company. That is intellectual property too, admittedly a different, not legally, but more emotionally based form. So, every time these experts go out and participate at a scientific conference, a trade show, a regulatory association, an expert group meeting, and every time they patent or publish their findings of their most important and exciting project work, they do represent the company and can add, or subtract from the company's value and can be defined as an element of IP.

For example, my former company was never considered to be part of the top 50 most innovative companies, formerly listed by *Business Week*. Some of us in the company worked with journalists of that journal to demonstrate the innovation power, capacity, and drive of my former company. Ultimately, we must have been convincing because the company, for the first time, appeared, as I believe as No. 38 on that list and the following year became even No. 36. Unfortunately, this type of listing was eventually discontinued and replaced by other types of listings; however, it shows how important the individual is, even in a large organization, and how he or she can be a strong representative and express the IP value of the company in substantial and successful ways.

4.1.6 Alliances and partnerships

Another important and often overlooked building block of IP is the entire area of alliances and partnerships, strategic or tactical. These can add tremendous value to the company, to its knowledge base, manufacturing know-how, new consumer groups, and markets. Alliances or partnerships are great vehicles to create know-how or to tap into existing know-how, thereby speeding up any development process. Moreover it binds the partners in defined and controlled ways. As far as I can judge, such partnerships are not really taken into consideration when financial analysts establish the value of a company when calculating and adding up all elements of the company's IP portfolio.

Figure 4.1 shows an overview of the typical elements of IP in the food industry.

4.1.7 Protect everything!

It so happens that companies keep their secrets too well and especially too long, either by not protecting them or by not sharing them at the appropriate time. Let me give you two examples for what I mean and which I have experienced in my former company. They are examples and could happen in any company. The mantra of the company was and largely still is to protect and own every technology, process, sometimes even ingredients, and any other ownable

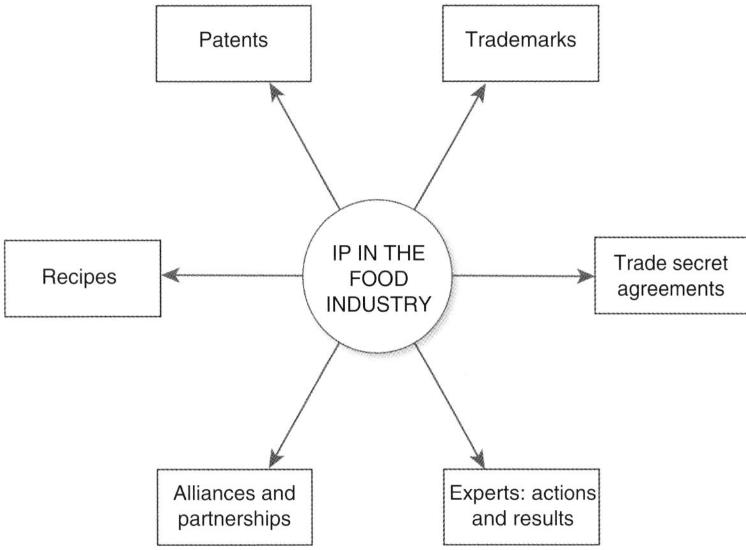

Figure 4.1 A holistic overview of IP in the food industry.

intellectual property. Although this is not a bad approach, it can lead to undesirable outliers and outright counterproductive results.

The first example is a fairly old one and happened in the area of soluble freeze-dried coffee. No worries, I will not (better: simply cannot) divulge any secret (!) information here. The company had developed the revolutionary continuous freeze-drying process, which dates back more than 50 years! It was really a great breakthrough. It was an enormous technological step forward and had an important impact on the taste and flavor profile of the resulting product, the new freeze-dried soluble coffee.

Needless to say, that not only was the company, its inventors, and all involved proud of the achievement but they were also secretive about the entire process. The company internal engineering groups had made great contributions and had all the drawings necessary to build the machine and all that had to come with it. A local machine builder was commissioned to build the equipment in the required size, several times over for all the factories that were supposed to manufacture this product. The equipment was so robust that its life cycle was so long that for many, many years it was not necessary to build many new machines, if not the one or the other to satisfy the needs of further geographic expansion of the product line. It was clear that the manufacturer of the equipment could not survive with this rate of business and turned to build other equipment and over time lost all knowledge of building the continuous freeze-drying equipment.

The real story actually starts right now. After 25 or so years, the time came to replace the original equipment, only to realize that the knowledge was lost, the equipment builder was not capable of building this particular equipment any longer and because the company had kept everything secret and highly confidential and did not license out the technology, no other machine-building company was in the position to build such a machine. However, there was interest from other, competing coffee companies out there, which forced the machine builder to come up with alternative freeze-drying technologies, and therefore developed the "batch freeze-dryers.

From an engineering point of view and from a factory operator point of view, this batch technology is not only less elegant but also more complex. So the inventors of the continuous freeze-drying technology had to bite the bullet and acquire batch freeze-dryers in the open market. Well, given the size and complexity of the technology, this open market is small. Anyway, the lesson to learn from this is simple: exaggerated secrecy and holding back of proprietary know-how for too long a period is never a good thing. Had the company licensed out the continuous freeze-drying technology after a reasonable period of time, say 10 years or so to machine builders, it would still have had a substantial time advantage over competition but could have negotiated advantageous conditions for the purchase of future machines as a result of having given the license. Be that as it may, as I have said time and again, hindsight is the only exact science and now we know better.

4.1.8 One last attempt

Let me close this story with a short epilogue. Approximately 30 years after the first implementation of the continuous freeze-drying technology, the company wanted to revive the technology inside its R&D group, calling on the "last knowledge carrier," who happened to still work in the same R&D group. Some of the original pilot-size equipment was still around, much of it in the "chunk yard" of the R&D center. Suffice it to say that the effort was rather futile and was given up some 9 months later with the recognition that too much time had passed, and it was simpler and safer to purchase necessary new equipment in the marketplace.

By the way, this is not the only example of missed opportunities in the context of attempting to keep everything not only secret inside the company but also not even protecting it through patents. Not wanting to delve into another story with a similar core message, I remember having sat in a kind of "panic meeting" when we found out that our jealously kept process for extracting coffee as one step in the entire process of manufacturing soluble coffee was patented, in almost identical ways to ours by the largest competitor. This was the moment when all lab journals, technical notes, and experts had to be consulted and shared with the judges in court to settle the matter, more or less, once and for all. Must have been a great satisfaction to learn that they were totally barking up the right tree.

Let me just say that the stories I just used as examples for missed opportunities because of badly interpreted and applied approach to handle IP is typical of larger companies. Smaller companies, and especially smaller food companies, would most likely cover themselves in different and more pragmatic ways, especially by working through partnerships with larger corporations. I am not aware of examples similar to the these in smaller food companies.

4.2 THE VALUE OF INTELLECTUAL PROPERTY FOR A FOOD COMPANY

Protect at any rate! That could be a fair description of the mantra of companies, especially large food companies. The position has changed a bit in the more recent past, but let me explain with a bit of a historic perspective, pretty much in sync with the overall title of this first part of the book. I vividly remember that one of the first slogans that one of the former CEOs of my old company came up with, even before he started in his function, was something that went like this: Every effort in the R&D world of the company should lead to a result that is:

- proprietary
- protectable
- patentable

Before I even have to say it, you have certainly already discovered that the initiative that resulted from the slogan, this message was: PPP.

4.2.1 Poor principles in practice

The good thing behind this was that at least everyone in R&D got a clear direction what he or she had to do and what was to be achieved; however, this narrowed the playing field enormously. For some it was an outright asphyxiation of their activities and a preferred excuse as to why certain topics would not be pursued any further: not patentable; we cannot sufficiently protect it; or somebody else owns it and would share it. PPP became the great dominator in the entire R&D group and closed the view to many great things that could have come, hadn't this darn PPP stopped us. Yes, I agree, it was and is an excuse, but used it was as an excuse extensively.

My critique here is that such extreme push for owning every possible IP can almost become counterproductive, although the opposite was and is intended. It was certainly well meant, but then, what is well known is valid here too: the opposite of good is well meant. One the one hand, this extreme push potentially creates a valuable portfolio of IP; on the other hand, it stymies much-needed drive for and achievement of innovation and new product, process, and service solutions for the company. Search for new innovation becomes inflexible and follows, or rather followed this rule of PPP. As in every community, some people liked it others hated it, but all had to live with it.

4.2.2 Change is on its way

Over the years, however, a kind of silent revolution happened and led to a much more flexible and pragmatic approach "owning of intellectual property." In Fo*od Industry Design, Technology and Innovation*, I briefly discussed a successful way forward when it comes to dealing with ownership of IP, "I remember times when I had to negotiate for 18 months with a university that did not want to budge over the question of who would become the owner of the IP of an as-yet non-existent collaboration." I described the situation of collaborating with an external expert resource, in this case, a university.

> *There was, however, a very important lesson from all this: there must be another, simpler, more successful way of dealing with such IP questions. When developing the innovation partnership model, I also developed a new approach to sharing IP between two partners, in actual fact a simple*

one, based on giving up the formerly iron-clad position of "we must own everything." Once this entered the mindset of my colleagues, things came together very quickly. What do I mean by this? It's very simple and it's got to do with the definition of owning. The mindset change was from, for instance, owning a patent to owning the application of such a patent. Let me give you an example: a flavor ingredient partner jointly develops a set of molecules by smart extraction, which is designed to substantially improve the taste profile of our products. Now, we progressed in very simple and pragmatic ways, by defining the ownership: the flavor partner owns the set of molecules and a patent that protects the know-how, but also defines the different possible fields of application. The food company, on behalf of which the development was performed in the first place, and which has also put in-kind resources into this development, subsequently receives the following freedoms:

- *Exclusive, license-free application of the said set of molecules,*
- *In a negotiated number of product areas,*
- *For a negotiated geography, (i.e. in which markets, and*
- *For a negotiated period of time, typically +/− 3 years.*

The duration is normally volume dependent, i.e. duration can increase with increasing volumes, but should in all cases be finite, simply because through selling the set of molecules to third parties, the future price of ingredients will become more affordable because of higher overall volumes" (Traitler, Coleman, & Hoffmann 2014, pp. 230–231).

I can assure the reader that this then novel approach ("then" meaning only a few years ago) was an eye-opener for all of us involved and made the playing field of negotiations with expert partners of any kind dramatically easier. There may, of course be reasons why this approach may not work for your company or is not indicated because of critical core know-how that needs to be protected solely in the name of the company. However, it is important to realize that in most cases where proprietary ownership of IP is suggested the only way forward, it is not really necessary and with a little bit of thinking and differentiated views on the individual cases and situations, the entire IP process could be rendered much more efficient and easier to handle.

It is good to see that the industry is apparently becoming more pragmatic in this subject matter and follows more and more the approach that I have outlined, which makes the life of everyone involved definitely easier and can help accelerate the process of generating and using IP in important strides, thus, once again, bringing any related product, process or service solution to market much quicker.

4.2.3 Patents forever

This brings me to another topic that discusses the emphasis and almost obsession of a company and especially its R&D group with patents and their use in value creation. Scientists and engineers alike have a big, common hobby: they like to talk about their work, more so they like to write about their work (and I know what I am talking and writing about…). It is obviously part of their reason to be the salary you receive at the end of the month will be more or less gone by next month's end; however, the written word of a publication, an internal memo, an internal R&D report, the text of a speech given, a chapter written in a book, and, last but not least, a patent is here to stay, at least for a good while. In most instances, it's all that's left of you, and some scientists and engineers see this as their main, if not only legacy.

This spirit of "making one's mark" together with a more or less gentle or forceful push by the company to protect the work of its scientists and engineers makes for a fertile playground to push toward as much IP as possible. As mentioned before, financial analysts do look into the number of patents, as well as other IP elements, to evaluate present and forecast future value of companies together with other elements such as the obvious revenues and profit margins and more indirectly compounded values of their brands and other assets. So, patents do play an important role in such value evaluations, and it is mostly numbers of overall patents applied for per year. This seems to be surprising because not every patent application leads to a granted patent. I was personally active in obtaining patents for my former company; over the years I came to some 25 international patents (not 1 patent in 25 countries but 25 patents, almost all of them as worldwide as could be). I recently (early 2015) added a granted patent in the context of my own startup company and was understandably proud of having done it on my own.

4.2.4 Numbers and more numbers

By writing all this you can immediately gather that, I, too, am proud of numbers and content and usage of patents and the value of outcome based on such patents. As stated, there are other, probably more important value contributors than the sheer number of patents but then numbers are easier to calculate and assess than content and impact. So, let's get to these numbers in a bit more detail. I paint the following picture regarding number of patents (not differentiated between applications and granted) per company (industry dependent) and per year shows the following data and estimates.

The large food companies such as Nestlé, Unilever, Kraft, Coca Cola, PepsiCo, Kellogg's, Mars, Mondelez, and a few others are in no way leaders in any patent statistics, and compared to most other industries, irrespective of country where patents are finally assigned, they are almost insignificant as far as numbers of assigned patents are concerned. In US patent statistics, there is not a single food or beverage company to be found among the top 10, which since 1997, and until 2014 was and is always headed by the IBM Company (6788 in 2014), followed in 2014 by Samsung (4652), Canon (3820), Sony (3073), Microsoft (2659), Matsushita (2582), Toshiba (2365), Qualcomm (2103), LG (1945) and Google (1851) ("List of Top United States Patent Recipients).

The following source from the US Patent Office (nd) shows the entire list for 2012. The first food-related company on this list is Procter & Gamble (P&G), although its food part has become small, with 395 granted patents in 2012. Nestec SA shows 72 patents to her name. It has to be said that the large number of Nestlé patents is first applied for in Switzerland, which probably makes the final number of granted patents per year a bit higher but not by much because ultimately almost all Nestec patents will also be filed in the United States, often at the same filing date.

4.2.5 And more numbers

The list from the European Patent Office in 2014 looks as follows: Siemens (757), Samsung (712), Robert Bosch (633), LG Group (630), LM Ericsson (574), BASF (543), Huawei (493), Panasonic (447), General Electric (434), and Qualcomm (427). This statistic shows another interesting number: the success rate of obtaining granted patents from the total number of applied

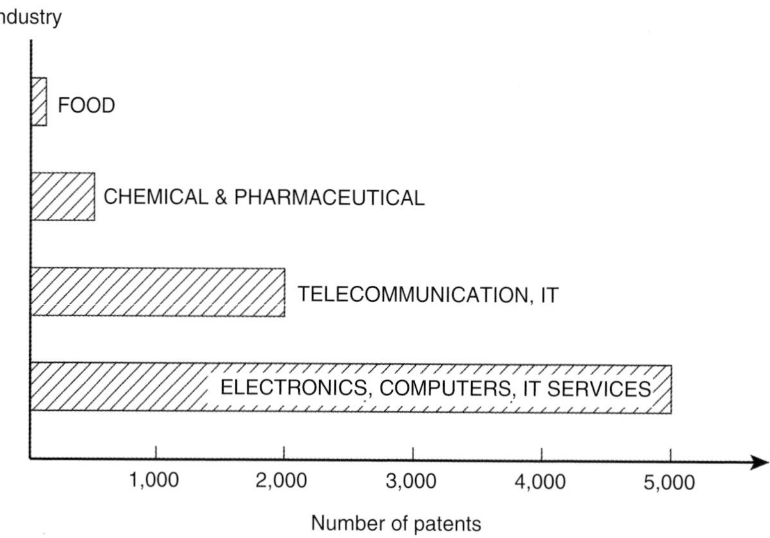

Figure 4.2 Average number of granted patents by industry in 2014.

for patents is pretty consistently 30 percent, or in other words, one of three applications gets accepted and granted, most likely after some modifications of examples and slimming of claims.

From the best of my recollection, a food company such as Nestlé applies for approximately 250 patents per year, and the numbers for comparable food and beverage companies are probably in the same order of magnitude and smaller, as can be seen from the US list of granted patents in 2012. Applying the success rate percentage of 30 percent, this nicely confirms the 72 patents granted to Nestec SA in the United States in 2012. The only other food company to be found on the 2012 list is Kraft Foods with 50 granted patents.

Figure 4.2 depicts a simplified overview on the average number of granted patents for different industries for the year 2014.

4.2.6 Here are more and even bigger numbers

Not wanting to bore you too much with statistical data, let me finish with more relevant information regarding the evolution of different patent areas over time. The observed time period here is for filing years from 2001 until 2005 and shows several important trends. The most important increase with an overall 4.9 percent in 5 years can be seen in the area of semiconductors, already beginning at a high absolute overall number of patents of more than 422,000 for the same period. The biggest loser is the area of IT methods for management with –9.8 percent and a total of approximately 125,000 patents.

The food industry (represented by food chemistry in this statistic) shows a good growth of 3.7 percent, representing an above average growth compared to the many other sectors in this set of data. The overall total number of filed patents in this area was 117,000, so not too shabby at all.

Given that typically only one-third of the filings are also granted still shows a total of close to 40,000 patents over 5 years or something in the order of 8,000 food-(chemistry-)related patents worldwide. There certainly are food-related patents in other areas listed in this statistic, like "electrical machinery, apparatus and energy—measurement instruments, analysis of biological materials, control—biotechnology, chemical engineering—machine tools, engines, pumps, thermal processes and apparatus." Without going into each of these areas in detail, it is virtually impossible to judge what percentage of patents in these areas is food-related but from the numbers involved, I would assume that the total amount of food-related patents in these other areas could be similar in numbers to the 8,000 patents which can be attributed directly. Calculating the average total number of patents for these other areas amounts to approximately 400,000 per year. Making the daring and maybe not permitted assumption that 5 percent of these patents are food-related in one way or another would add another 20,000 filed or approximately 7,000 granted patents per year for the filing years 2001 to 2005. This could lead to the, again to uncertain assumption that the food and related industry in total, worldwide obtains some 15,000 granted patents per year. Actually, this looks like an impressive number (World Intellectual Property Organization 2008).

4.2.7 Is my patent actually profitable?

The really important question that has to be asked now is: "How many of these patents that hundreds if not thousands of companies, universities, other organizations, and individuals file and obtain granted are actually useful for the assignee, and how many of these granted patents are actually used for any given length of time?"

I can say with some authority that only a small fraction of all granted patents are actually really at the core of what any company uses to protect its products, its processes, its services, and ultimately its brands. One may disagree on the exact numbers but I suggest the following split:

- Core patents: ~15–20 percent of total patent portfolio
- First line of protection, the safety perimeter around core: ~ 20–30 percent
- Other, not directly necessary to protect core and partly unrelated: ~ 50–65 percent

Let me briefly explain the three classes.

Core patents are really those, which are at the heart of the company's product, process- and service-related assets, and therefore most likely necessary to be entirely and solely owned by the company. These patents are active and are being used all the time. Companies make big strides to defend these and to find extensions to these through new and related inventions that can be patented to continue protection and defend its assets.

Figure 4.3 illustrates a typical patent fortress architecture.

The first line of defense represents the type of patents that are close to the core patents and can hold off a competitor to such a degree that they cannot approach the IP of the core patents and endanger it.

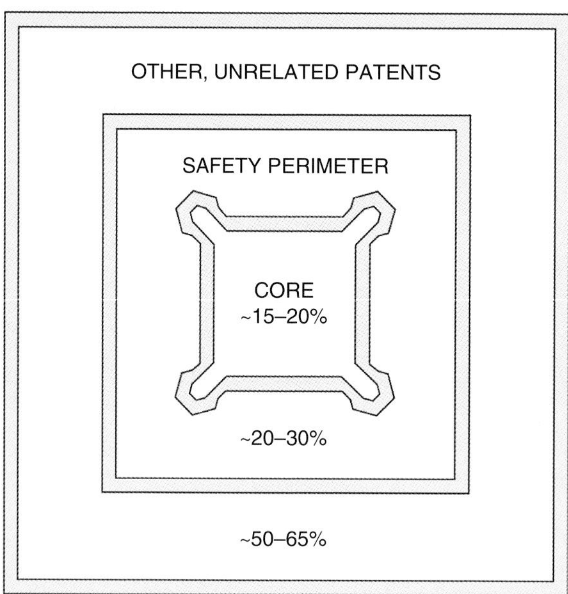

Figure 4.3 A typical patent fortress in the food industry.

The large majority of patents, however, are those, which do not link either to "core" or "safety perimeter" patents, and is mainly the result of the creativity and innovativeness of the company's scientists and engineers or other creative minds, including designers and procurement, just to name two. Such patents represent the intangible value of IP and can often be the starting point of a new product line, a surprising new technology or a new level of service that the company can offer to its customers and consumers. At first sight, such patents do not yet present or create tangible value but almost act like sleeper cells; they are here, they don't do anything, yet but one day they might surprise us. For the better, that is.

4.2.8 It's all about brands! And about service level!

In the end, the real value of IP is reflected in the value of any company's brands, in cases that we are looking at a branded company. But even for the so-called no-brand companies there are brands, store brands at the end of the manufacturing line and distribution chain, and I argue that we have here the same situation as for branded companies: the entire value of know-how of IP that is created and used, of "savoir-faire" is reflected in the brands of any company and in its level of service that comes with these brands. This also means that brands and services unite what I would call collective wisdom of any company where technical IP meets marketing artistry, sales shrewdness, and procurement smartness to form a strong bond, almost a super IP, an IP^2, which in its entirety represent the real and total value of any company. Technical IP alone is only one element of this, although not too small an element.

4.2.9 Good communication is key, great communication creates value

Let me close this section on value of IP, with the intangible value of interactions that every company, and especially its R&D representatives, has with industry organizations, consumer groups, government and nongovernment organizations, with the media, and finally with the outside world at large through understandable and *honest* communication. I emphasize honest, not because I suggest that companies would not be honest when it comes to communicating with or to the outside world (outside the own company, that is), but because I rather suggest that companies don't do this enough and often only in times of a crisis and to defend themselves after the fact. An important part of good communications is telling the truth; the essential part of great communication is to anticipate possible events or issues and handle them in a truthful and responsible manner. For a consumer goods company, such as a food company, this means to always communicate with the consumer in mind.

This often means to speak to consumer groups, discuss with NGOs, bring critical and difficult issues to the fore in industry association meetings and, last but not least, follow quality and safety rules and regulations of every individual market (country) to the letter and not getting caught for violations, even minor ones. There have been cases in the food industry where traceability was clearly not sufficient, or where there was ambiguity in interpreting local regulations, both of which lead to subsequent court actions. Every time this happens, the industry at large is almost taken hostage in the public opinion and its credibility and reputation suffer considerably. Yes, the public tends to forget with time, but this is still no excuse to not follow the rules. I discuss this here under the heading of IP because wrong actions in this field of quality, safety, and regulations not only taint the image of the food company in question and to some degree the industry at large but also because such wrong-doing is clearly a destroyer of value that was otherwise created by innovative and sophisticated IP elsewhere in the company.

4.3 INTELLECTUAL PROPERTY AS THE BASIS FOR INDUSTRIAL INTELLIGENCE AND COUNTERINTELLIGENCE

It is not my intention to tell a spy story here but rather to discuss a clever reading of IP to discover which directions competitors in the field have taken or will take, or more surprisingly, which your own company might head toward and which appears not to be in sync with your company's published strategy. Let me begin the discussion and analysis here by discussing the nature and value of an IP portfolio, which includes such items as patent portfolio, technology portfolio, more generally know-how portfolios, by identifying the carriers of such know-how, the experts. There are a few more members in this portfolio club, for instance the portfolio of external alliances or innovation partnerships to name just these two.

4.3.1 List everything

I have personally gone through several waves of attempts to establish a list of key technologies in my former company, and it always went by the same playbook:

- Management asks the technical group, typically R&D, to establish a list of key technologies
- R&D experts debate what *key technology* means and which technologies should be part of such a list.
- The debate becomes heated but eventually, the question reaches the lower echelons in R&D.
- These echelons are asked to list their key technologies in their respective field; to make things more complicated, they are also asked to give a short description of every technology they list, and especially where the technology is used and for what purpose and with which success.
- Tons of data come back to the higher echelons, and they immediately see (they are not dumb, that's why they are the higher echelons…) that what they have received back is far too complex and especially far too long.
- The reader may recall that the question as to what is a key technology in the second bullet point remained unanswered and management now, on the basis of the returned lists, debates this question.
- In most of the cases that I have experienced, there was never a clear answer to this; hence most of the time, the action stopped right there. Maybe the excessively long lists were internally published more or less as such and probably reflected a fairly good image of the company's technical capabilities and one could even give such a list the name of a technology portfolio, although it rather was a shopping list without much differentiation.

4.3.2 Technologies and people

As indicated, I have seen such exercises in almost identical ways happen several times over the years, and maybe it's just a reflection of the difficulty of the task at hand, especially for larger food companies. Additionally, there is a strong human element involved in all this: the individual carrier of expertise that represents a specific technology would not want to have to admit that his or her topic is not key but just supporting. That's almost like having to admit that what I do is not important. No way! Therefore, almost every piece of know-how and expertise was to be found on such lists of key technologies. This is also valid for the question what the company's key competencies are, which is a slightly different question than the one presented previously. Although *key technology* is more or less impersonal, the question related to key competence becomes personal, or linked to a person or a group of persons, the experts.

4.3.3 Who are the experts?

The latter most often leads to conflicts inside the organization: do we want to or need to list our experts' names and corresponding competencies? What if such a list would fall into the hands of headhunters or worse, competition? I have seen several of these initiatives, or rather quests to establish such lists of people and competencies, and I have seen these lists. Maybe it's simply paranoia or some milder form of it, but I still believe that with today's means of IT intelligence, such a list could quickly become public knowledge and fall into the wrong hands. On the other hand, and my "other soul" speaks now, competition and headhunters would have and do have ways and means to find out what a company is doing and what competences it possesses and who are the carriers of these competencies. How? Well, simply by following patent literature, speeches given by scientists or engineers at conferences or trade shows, scientific publications, and so on.

Not much remains hidden, and it is relatively easy to obtain a good picture of what a competitor is capable of and who stands behind these capabilities. After having participated in many discussions and deliberations regarding this issue of competencies and names, I must say that today I am all in favor of such lists, because they most and foremost serve the purpose of communication within your own company. I have lived many situations in which management really had no idea, and when I say no idea I really mean *no idea* of what certain if not many people did and worked on and were competent in within their own R&D organization. I do not suggest that this happens in your organization or that it even happens today with all the IT and administrative support that is given to establish such lists and follow everyone ever so closely. On the other hand all information is originating from the individuals themselves and with increasing administrative demands, people have a tendency to become a bit superficial and brush over details so not to lose too much time with such trivialities as filling in lists, which, on authority, not many if any read at the end of the day.

4.3.4 Don't ask questions, just fill in the form!

Be this as it may, it all depends what your R&D organization wants to achieve: properly filled in lists of detailed descriptions of your hours and days, hours spent, competencies needed, split of business for which the work was intended or, alternatively, solid, sustainable, and profitable results? This is obviously a leading question and the logical answer has to be: solid, sustainable, and profitable results yet, in most instances I have experienced such logic was defied; hierarchy mostly wants control and wants your time sheet, in whichever way it comes today, filled in properly. This may sound strange to you but it isn't so strange anymore knowing the self-fulfilling prophecy of complex and all-encompassing IT management systems that need to be fed with data, numbers, keywords, hours, acronyms, short descriptions, more numbers, and more short descriptions, and so on. You get the picture? It's almost like the proverbial Moloch that devours its children. I know for a fact, or let me rather say, I knew for a fact that the only people reading properly through such compiled data sheets were the internal auditors, mainly to find out whether something was not done according to the agreed-on auditing criteria and subsequently suggesting appropriate and corrective action; to correct what? Well, to follow rules even better and fill in those pages even smarter and according to the set rules. It's a real pity because a much simplified version of these time or rather activity sheets could actually be really useful for everyone involved; the expert who has to summarize his or her work in the shortest yet most explanatory ways, the people who are on the receiving end to learn about progress, results, and promises, and the system at large, which would have short and concise documentation to store and be able to consult at later stages whenever needed and for whichever purpose. I have seen attempts to achieve this by requests to write "highlights" of one's work, for instance monthly. However, there was never a real definition of what a highlight was supposed to be and so people started to write fairly lengthy status reports of their ongoing work without selection and discrimination as to what was a real highlight of the month and worthy to be reported on.

4.3.5 I want monthly highlights, although I don't read them

This led to the result that these monthly highlights became heavy, unreadable documents with the main result that much time had to be put aside to write them for the sad purpose that no one really read them; they were just there, and yes, they could be consulted if one wanted but there was no real structure to these monthly highlights other than being voluminous and therefore unusable. Compilations of really well-made monthly (or quarterly, or biannually) highlights could be a rich source of easily retrievable intellectual property of tremendous value for the R&D organization and ultimately for the company. Sounds like a lost opportunity or an occasion to invent something new, a new reporting system, ideally no system at all.

I had a friend who was a painter and who had worked at the Disney Imagineering, the really creative branch of the Walt Disney Company. Employees of the Imagineering were required to fill in time sheets, but they could take certain liberties as to how they were executing this requirement. My friend, being an artist, came up with a studious and creative solution: he simply drew his time sheets; at the end of the day he took a few minutes and created a short cartoon outlining his daily activities. Needless to say, that at the end of a month or a year he had a really impressive collection of sketches and drawings in cartoon format, nicely and understandably, depicting his activities during any given time period. The part about "understandable" was probably the best and most rewarding one of his approach to filling in time sheets.

I do not suggest that everyone in R&D should become an artist and draw his or her time sheets or comparable activity reports, but you might want to try it out or find you personal way of helping the system to leave the system behind itself. This would also be an efficient way to make counterintelligence more solid, and it would make it more difficult for competitors to discover the competencies of individuals in your organization and in turn hinder competitors to easily discover the strategic and tactical intent of your company. Generally speaking: the less systemic your approach to intellectual property, the more difficult to penetrate the fortress for someone from the outside. I admit that it makes it possibly more difficult to understand also for your own management, but at the end of the day, management is paid to understand and not paid for making everything easy for them. Wouldn't it be so much nicer if Gantt charts were to be replaced by fun cartoons and balance sheets by Sudoku charts? I know that this is not going to happen soon (or ever) but one never knows.

4.3.6 Open up!

Because everything is so transparent and clear, management is afraid to open up to the outside world, although every company has a more or less strong commitment to "open innovation."

Open innovation is largely described elsewhere, for instance in the prominent books and writings of Henry Chesbrough. Open innovation became popular in the early 2000s, it's best-known example being P&G. The example is so well known because it also was at the onset of the widely known "connect and develop" (C&D) initiative that was again widely written about in many books and publications. Allow me one important observation here: in the case of P&G and C&D, the approach of open innovation was almost a feat of despair because the company was not doing so well in these days (end 1990s and early 2000s) and something had to be done. Most companies, without wanting to be cynical, would "go back to basics," while P&G courageously chose to try something new, namely open innovation, C&D and opening up to the outside world of experts and competencies. They were not afraid to divulge their strategic intent when they openly asked for solutions to problems, rather the proverbial opportunities, for which they did not have solutions, at least not fast enough.

When I had the chance to establish the approach to open innovation in my former company, we had to work around this aspect of opening up too much. Some of the top managers had the fear that by doing so (i.e., opening up and asking for outside help), we would give away not only our strategic intent but also stakes from the value chain. Although there might be some truth to this, the major benefits from open innovation, any type of open innovation that is, comes from increased speed of executing a project faster and being in the marketplace potentially twice as fast as one could without "professional help" from the outside. We solved the fear of indiscriminately opening up and potentially losing out by making the approach to open innovation a fairly controlled partnership approach (i.e., choosing reliable, trusted, and competent partners to become members of a carefully selected group of highly rated innovation partners). This tempered the fears of the fearful and pleased those already convinced of the need for opening up to the outside world.

I have written quite extensively about this approach in *Food Industry Design, Technology and Innovation* (Traitler et al., 2014) and encourage the reader to consult the chapter on open innovation and innovation partnerships in more detail.

4.4 COMMERCIALIZING IP ASSETS

Let me introduce this last topic of this chapter with one short sentence: commercializing IP in the food industry is not popular at all. And this is almost a euphemism: it's not popular, it's hardly used ever, and there are some examples to the contrary, which are at the borderline of food-related topics and competencies to out-license and make money with. Why is that so? Difficult to say but let me try it nevertheless.

Let me begin with the company I know probably the closest. The answer to the question how much did Nestlé earn with out-licensing technologies, ingredients, or products is close to an insignificant amount. I have to be clear that I do not include royalty makers such as KitKat® or Rolo® in this assumption. Nestlé receives a certain percentage of royalties for these two products from Hershey's for the fact that Hershey's had negotiated a license deal for the US market with Rowntree more or less in the year before Nestlé acquired Rowntree Macintosh back in 1987 or so. Approximately one-third of the total global production and sales of especially

KitKat® and a lesser amount for Rolo® is performed and achieved by Hershey's, which brings a fairly nice chunk of royalties into the Nestlé coffers. Not enough say some and tried, in vain, to back-acquire the brands from Hershey's a few years ago.

4.4.1 A good license deal is better than no license deal or so you would think

To the best of my recollection, there is not much else in money made by the Nestlé Company in exchange for out-licensing activities. I have to add though that out-licensing is not part of Nestlé's business model, which strives, still to this day, to cover as much as possible of the entire value chain from field to fork, yet it is not vertically integrated. The company does not own a single crop field, if it was not some small experimental acreage. It was tried a few times to gradually change this approach, but none of these attempts were successful and therefore not forcefully enough pursued or retried. That is not to say that through increasingly opening up to complementary knowledge providers the attitude and approach toward outsourcing jointly developed technologies or ingredients could not become a new direction for Nestlé. A minor, first example for this is the new approach to "owning" intellectual property as described and discussed previously in this chapter.

Let me present another example for out-licensing at P&G. When they started the well-known and largely elsewhere described connect and develop (C&D) initiative, it was apparently understood from the beginning that the fairly heavy investment into C&D had to be covered, at least in part by outsourcing of technologies or know-how that was neither critical nor strategically used by the company. C&D meant to establish a large group of technical experts to assess incoming ideas, as well as a back office with lawyers and other personnel, which in its hey days accounted for some 130 people worldwide. This had to be paid and what better way than to use the same people, especially the lawyers, to negotiate valuable and profitable license contracts with interested external partners. And so they did. From many personal discussions with former colleagues at P&G, the order of magnitude of value achieved for out-licensing was quickly rising to approximately $2 billion, certainly a highly desirable amount, which largely paid for the C&D effort, at least that's what it looked like.

4.4.2 Licensing out most often is a deviation of the traditional business model of a food company

I have to add that the interest for open innovation as well as the creation of C&D was largely driven by a rather underperforming business, and the desire, if not need, to try out something radically different from what was done before: "If you do what you always did, you will get what you always got," or, alternatively, doing something differently, you will obtain, most of the times, different, likely better and more successful results. I am not advocating that companies should go through bad and unsuccessful phases to dare new and unchartered approaches, but there is something to this when disruptive approaches are required to save businesses.

There are not many other examples for out-licensing in the food industry and the P&G example is actually not a purely food-related one. To the best of my knowledge, most of the out-licensing income of P&G came for other, non-food parts of the business. Out-licensing and making money with one's own technologies and know-how does not seem to be part of the genes of food companies, unlike many other technology-driven companies, be it computers, software, automotive, or even the world of space exploration. There is a saying which goes like this: "Whatever new and apparently disruptive technology or even gadget you find in today's Mercedes S-class, you will find in most other cars in 10 years from now." There is much truth to it, even if you could replace "S-class" by other high tech automotive brands, such as Porsche or others.

Let me terminate this chapter with one example of out-licensing of know-how in which I am personally involved. The Jet Propulsion Laboratory/NASA (JPL) of Pasadena has begun a program of out-licensing mission-noncritical know-how, mainly technologies to interested partners, for instance, the food industry. They describe this by the term *reimbursables*, clearly indicating that they have understood the signs of the time of opening up and making their know-how available to a much larger and not necessarily directly linked client group than they ever were dealing with. If you are interested in learning more about this JPL initiative, here is the link: http://scienceandtechnology.jpl.nasa.gov/opportunities/

4.5 SUMMARY AND MAJOR LEARNING

- This chapter was all about intellectual property (IP), how it is handled in the food industry, how it is valued, and the elements that fall under the large term IP.
- IP is not only just patents; IP is patents; preliminary invention disclosures (including information recorded in lab journals or similar IT based formats); provisional patent applications; recipes; trademarks; trade secrets and secrecy agreements; nondisclosure agreements; and finally the company's own experts, their actions and alliances and partnerships with third parties.
- In the past, most food companies, especially the larger ones, attempted to protect any kind of IP in their own name, thereby hoping to maximize its profiting from the entire value chain.
- This trend to keep everything for oneself as much as possible has unfortunately led to expensive protection mechanisms, and more so to lost opportunities as described in the example of technology for continuous freeze-drying of soluble coffee.
- Because not everything was protected either solely or in partnership, sometimes competition patented exactly the same procedure used by the company. Although annoying, this did not mean that one's own technology could not be used anymore, but it had to be proven that the competitor's patent was duplicating the own approach and therefore obsolete.
- The value of IP for the company was discussed and especially the desire of the food company to protect everything as much as possible going as far as to only work on proprietary, protectable, and patentable topics.
- More recently, companies have become more pragmatic and rather want to own the freedom of usage of IP, its application, defined by parameters such as field of application (which products or product group), which geography (which market, country, region), and by expected volumes of sales, the latter determining the duration of such "application ownership" agreements.

- Value of IP is generally important for external valuation of companies by financial analysts and overall number of applied for and ultimately granted patents per fiscal year is still a prominent element of valuation, next to many others of course.
- It is estimated that in the food industry, like in many other industries, the success rate of granted patents is in the order of 30 to 35 percent of total patent applications. Absolute numbers of granted patents are much smaller for the food industry as compared to electronics or automotive companies.
- Total estimated numbers of granted patents in 2014 are as follows: Food ~150, Chemical and Pharmaceutical ~500, Telecommunications and IT ~2,000, Electronics, and Computers and IT Services ~5,000.
- The importance of the "patent fortress" was discussed and analyzed, and it was suggested that in the food industry: 15–20 percent of all granted patents are core, critical patents, 20–30 percent represent a first line of protection to the core patents, and 50–60 percent of all patents are other, not directly related and opportunistic patents.
- The role and value of IP as the basis for industrial intelligence and counterintelligence was analyzed and discussed. This encompasses the systematic approach to establishing listings of own know-how such as key technologies, expert lists, and the like. The pros and cons, especially as far as lists of experts is concerned, were discussed and some of the dangers were pointed out.
- Monthly highlights or something along the lines of reporting on project highlights in a regular (or irregular as-needed) basis is a great tool for management to learn about ongoing projects in their own environment. More often than not, such reports are abused to report on everything and therefore can become unintelligibly lengthy and therefore I have observed a trend toward such reports not being read.
- The increasing trend toward open innovation represents a new playing field for technology or know-how intelligence as well as counterintelligence but also a great introduction to creating more value around the company's know-how base.
- The potential value of out-licensing of know-how to interested third parties was discussed, and it was concluded that there is a great aversion to out-licensing know-how in the food industry, contrary to industries that are active in more technology-driven consumer spaces.

REFERENCES

European Patent Office, 2015, List of yearly patents, retrieved from http://www.epo.org/about-us/annual-reports-statistics/statistics.html.

"List of top United States patents recipients," 2015, retrieved from http://en.wikipedia.org/wiki/List_of_top_United_States_patent_recipients.

"Trade secrets," 2015, retrieved from http://en.wikipedia.org/wiki/Trade_secret.

Traitler, H, Coleman, B, Hofmann, K 2014, *Food industry design, technology and innovation*, Wiley-Blackwell, Hoboken, NJ.

US Patent Office, 2013, List of yearly patents, http://www.uspto.gov/web/offices/ac/ido/oeip/taf/topo_12.htm#PartB

World Intellectual Property Organization, 2008, "World patent report: A statistical review," retrieved from http://www.wipo.int/ipstats/en/statistics/patents/wipo_pub_931.html#f1.

Part 2

Possible future of the food industry

5 The need for a new approach to R&D in the food industry

You can make a lot of mistakes and still recover if you run an efficient operation. Or you can be brilliant and still go out of business if you're too inefficient.

Sam Walton

5.1 R&D IN THE FOOD INDUSTRY IS INEFFICIENT: AN ANALYSIS

It has been mentioned before that high-ranking officers in the food industry (and possibly in other industries as well) love to say that they believe that their R&D departments cost too much, twice as much as they should actually cost, but that they don't know which half to cut. Not only is this a cynical statement, but it is also proof of poor judgment. High-ranking officers, especially CEOs, are paid to exert good judgment. Now I am running the danger of becoming cynical and that's certainly not what I intend to do. Let me attempt to remain factual, as much as this is possible in a field that is filled with "financial emotions" and "romantic ideas" of innovation and progress. The latter could be defined as overly ambitious expectations, the former rather the belief that research and development should cost as little as possible and especially innovation should be cost neutral or better yet, should cost less than the old innovation. Consequently, there is this strong belief that R&D costs too much, therefore it is inefficient at its base and should be financially strangled all the time.

5.1.1 Innovation at zero extra costs

Don't get me wrong, I do not pretend that restrictions are not a healthy approach to improve creativity and innovative output. On the contrary, I strongly believe in restrictions and constraints, especially when it comes to the basic structures and routine and repetitive elements of R&D such as administrative tracking and justifying work details. Therein lies a lot of inefficiency, and it's got a lot to do with this cynical approach, paired by mistrust of top

Food Industry R&D: A New Approach, First Edition. Helmut Traitler, Birgit Coleman and Adam Burbidge.

management toward R&D. Some of this mistrust is explainable by past experience of such managers with regards to the results, or non-results delivered by the R&D organizations; however, the larger portions of this mistrust stem from not properly understanding how R&D and how scientific and technological development and discovery work or don't. Without wanting to fall into the cynicism trap, I would almost suggest that this shows a strong weakness on the part of top management: on the one hand employees in R&D are hired, trained, well paid, and let loose to discover and deliver, and on the other hand those who often have little knowledge and insights into the dealing of discovery decide how much this discovery should and will cost. It's pretty much like "Manchester United" or "FC Bayern Munich" were coached by Theresa May or Angela Merkel. Nobody would expect this and yet, in a company, and especially in a food company, that's what often or almost always happens. And nobody complains or at least not with a loud voice.

This chapter deals with the topic of the need for a new approach to R&D in the food industry, and in this section, I discuss the theme of R&D's inefficiency to deal with the challenges of today's food industry. I see the need to discuss this question of inefficiency at all levels and that includes top management of any company, including all the traditional roles of the executive board, the CEO, and the president of the board but also members of the administrative boards of food companies. This is not a "revenge discussion" as some may think but rather a holistic approach to discussing the inefficiencies of a food company at large, and which part its R&D organization plays and how this role, function, and actions can and must be improved. In *Food Industry Design, Technology and Innovation* (Traitler, Coleman, & Hoffmann 2014), I discuss possible organizations of food companies and have listed four possibilities:

- The product-centric company
- The design-centric company
- The consumer-, trade-, and operator-centric company
- The shareholder-centric company

5.1.2 Real changes are required

I have mainly discussed the question of which basic functions of the company should be inside or outside ("outsourced") the company in each of these four variants. Knowing, or rather, sensing that the discussion of an "R&D-centric" company variant would be leading too far away from the topics that I wanted to focus on and discuss in that book. This "R&D-centric" approach is being covered to some degree in this book, although R&D-centricity is not at its heart; it's really the question of how R&D has to evolve and develop in a changing consumer, customer, commercial and business environment, taking it for granted that the company at large has to change and evolve as well and not only by lip service or the increased usage of IT and the all so popular apps on mobile devices. These are often excuses deviating from the hard questions as to how the entire company has to embrace these changes.

I do admit that it is often difficult, if not impossible, to distinguish between trends and new business approaches that might just be a fad or something real. It takes guts and some good judgment to go forward with something that eventually might turn out to be a flop but alternatively could become a great hit. Marketing does this almost all the time and is, also almost always, given the means to go forward. I have rarely experienced situations in which planned marketing campaigns were reduced in their budgets, although this certainly happens. But it is rather black and white here: either it is done full steam (and budget) ahead or it is stopped. Not so in R&D! This is one of the biggest downfalls of any R&D organization: they want to do everything, a little bit of everything because everything is *interesting*! Here we go, I used the "i-word," my best hated word. When you hear the word *interesting* as a comment or reply to one of your presentations you can almost be assured that nothing good will come; it's mostly a polite way of saying "OK but don't bother me with this any more; next?"

5.1.3 Small is beautiful; large becomes inefficient

That's the real basic problem of R&D in the food industry as I know it: everything is interesting. And there is a basic and functional, even logical, reason for this: the people, the experts, and specialists who work in the R&D organization have certain specializations that they cannot change like a shirt every day. A marketing person is still a marketing person, although admittedly there are areas of specialization here as well but it's more a specialization toward one specific product group and a new orientation toward another one is almost always possible. Not so in R&D: a "dry foam specialist" (or any other highly specialized expert) cannot easily work on new methods of tempering chocolate masses, just to mention one example. It is difficult, hence there is a strong inflexibility to be observed in R&D organizations that is not so easy to overcome and that accounts for its inefficiency.

There is yet another argument that needs to be discussed here and that has got to do with the size of the organization. One can easily observe that organizations start to operate like the public sector with increasing size and tend to become bureaucratic over time. It invents rules, regulations, red tape, and all kinds of obstacles that seem to be necessary because of the apparent size, which I accuse of being one of the major hurdles toward increasing the efficiency of the organization. This is almost like the growing of the proverbial Moloch, which becomes an obstacle in itself: large, slow, devouring, and not flexible to say it nicely, yet a monster after all.

5.1.4 The good, the creative, and the productive

Still, the question of which half of an R&D organization is the good, the creative and productive half, and which is the unnecessary, even useless half is not answered. Yet, it may never really be possible to give a clear answer. If you write a manuscript, a publication, a letter, or something else, and you are told that your text is too long by a factor of two you won't go back to your text and strike every second word. You simply wouldn't do this, but you will, quite logically try to rewrite, reformulate, and render the text more concise, and if this is not enough, you will, heavy-heartedly leave bits and pieces out, hoping that your text still conveys the important message you were intending to convey in the first place. That's never an easy task and some results may become, to say it nicely, suboptimal and difficult to understand or even completely misleading.

5.1.5 What's wrong with R&D?

Cutting any R&D organization in half or even reducing by a modest percentage is a tricky and difficult undertaking. There are many questions involved in such a process, and the answers are often difficult to find. Let me suggest just a few questions or topics that have to be considered when attempting to streamline and optimize the R&D organization:

- Is the output of the R&D group perceived insufficient?
- Are results coming too late?
- Is the R&D group inflexible and not responding fast enough to changing needs?
- Does R&D support the business needs closely?
- Does R&D understand the business needs? Are the business needs understandable in the first place?
- Has R&D grown through acquisitions and there is overlap, but maybe just not discernible?
- Has your company gone through many rounds of cost cutting exercises perceived of being asymmetric, and R&D should finally pay a price too? Is symmetry the right solution?
- Should investments in, and support of, R&D happen anti-cyclically? In other words: when business is doing well, tone down investments into the mainstream, and invest in new and promising fields to be prepared for those times when business suffers?
- How do you tell people that have been hired for their high degree of specialization and expertise that their specialization and expertise is not any longer required and not make a stupid face!
- Are there alternative ways of maintaining a baseline expertise, "cook on a low flame" to minimize the losses? The company has, after all, invested quite heavily in every expert and specialist that was hired and internally trained to deliver results that may no longer be needed.

Figure 5.1 depicts a list of perceived flaws and proposed remedies to streamline and optimize the R&D organization.

5.1.6 I don't know which half to cut!

There are probably quite a few more relevant questions one could ask and issues that could be brought up in the context of finding, defining, and correcting inefficiencies in the R&D organization of a food company, probably of *any* company. Let me tackle the preceding list, but I cannot guarantee at all that all questions and topics will be answered and covered to the readers' satisfaction. It's a tough call after all. The major question is, and I do realize that it's posed in a general way: is every R&D organization, especially in the food industry by definition, a priori, branded inefficient? If you listen to CEOs of such companies and even to some of the executive board members of such companies, then you may get a simple answer: Yes. Often or almost always this answer is even undifferentiated as expressed in the introductory quote to Chapter 1: "I know that our R&D probably costs twice of what it could cost but I don't know, which half to cut."

PERCEIVED FLAW	PROPOSED REMEDY
Low perceived output	Improve perception
Results come too late	Increase speed
R&D is inflexible	Provide for change
Too far detached from business	Increase integration
No good understanding of business needs	Involve R&D more in business decisions
R&D grew unnecessarily through acquisitions	Reduce overlap
Perceived asymmetric cost cuttings	Fewer, more equitable cost cutting
Cyclical investment cycle approach	Go to anti-cyclical approach
High degree of specialization of personnel	Increase in-sourcing of external expertise
Too much superfluous personnel	Define and keep baseline (core) expertise

Figure 5.1 Optimizing the R&D Organization.

5.1.7 Let's eliminate every second word

You may argue that if not even the CEO knows how to judge the efficiency of the company's R&D organization, and has no clue how to improve it (other than to order the next round of budget cuts) how would we, the "regular folks" in the organization know how to improve and eliminate efficiencies?

It's tricky because, by definition, the CEO does not run R&D (with few exceptions) and is therefore not obliged to know how to improve. His or her almost natural, finance-driven instinct is to cut numbers. You remember example, "how to shorten a text"?

The CEO and most of his direct reports choose the route of the example of eliminating every second word of the text (some percent of budget cuts now, some percent in the next round), totally undifferentiated and also un-reflected. One may argue that CEOs and executive board members should know better because they are paid better. Far from the truth, I am afraid to say!

So, there is no help from above. On the other hand, it is crystal clear if you allow me this image that every organization has, by definition built-in inefficiencies. There exists a direct correlation between size and inefficiencies without going as far as to say that "small is beautiful" in all instances. But I have seen this many times and tend to compare large organizations to public and government structures. I am far from government bashing because one should not wish for "no government" and then be surprised to end up like Somalia. That is, however, not to say that private enterprises should allow for government-like bureaucracy-type of work, especially organizational approaches.

5.1.8 Let's do another budget cut

There are two possible remedies.

First, organize the large organization in many smaller, flexible, and agile entities, and second, restrict their access to large funds and resources. The latter could be seen as budget cuts, but it's the contrary of budget cuts. Let me explain. During my many years in the food industry I can report that I have seen and lived through many budget cuts; some of them probably were justified to shed fat and other nonproductive areas (this is not only people but mostly activities), many others rather pure actionist events. Yes, money could be saved on the surface but underneath it is difficult, if not impossible, to judge and estimate how much goodwill, especially with the consumers at the end is lost. You may argue that consumers don't see this. Wrong! Consumers today through their personal, social, and media networks pick almost everything up, and this most often in distorted and misinterpreted ways. On the other hand, we all know that the consumer is king (or rather queen), and therefore it is probably not a good idea to start arguing about their perceptions.

Better, though, if budget cuts are the game of the day, explain not only to your own employees but also to the consumers at large why the company does it and what the expected goal is. It's not good enough any longer to just satisfy the balance sheet, the shareholders, and financial analysts but ideally the field of satisfied stakeholders includes employees, customers (the trade), and consumers. Every company that intends to cut budgets, cut costs, or rather attempts to improve cost structure and improve margins needs to have a world-class communication strategy at hand and ready to explain to all involved, internally and externally. The times of arrogance and "we just impose it" are long gone, but some, especially large companies have still not understood this.

But why is this still the case and why do companies turn to measures like cutting budgets as their first remedy when attempting to improve margins and "take costs out"? Well, the simple answer is because they always did.

5.1.9 Innovation is key!

Interestingly, everyone talks about innovation and innovative products and services as the most important elements for sustained success of every company. But when push comes to shove, the natural reaction of those who are influenced by the balance sheet and that the product portfolio decides the old ways: when getting to the river, they build a bridge to cross it and don't think of other, creative, and innovative means to cross the proverbial river. One would think that this only happens in larger, mega-companies, but I have seen this equally frequently in small companies as well and this natural reflex of "let's cut costs" seems to be an endemic that has befallen the industry, still to this day.

I still owe the analysis and discussion as to why budget cuts are not automatically a new approach to restrict the employees of the company, especially the R&D group, and thereby enhance and improve the creative and innovative output? Part of the answer to this question is that cutting budgets is never intended to restrict the creative minds and thus make them even more creative but have the only goal to improve margin. The fallout of that could be creating restricted environments but the only restrictions that I have ever seen in such a context are cuts in overall R&D budgets without differentiation and carving out new R&D spaces—fields of innovation—and environments. The sole goal is margin improvement. While this in itself is never a bad thing, it distorts the view toward how to optimally instill a policy and atmosphere of healthy and creativity enhancing positively perceived constraints environment. People do respond positively to this, not all, but most, and it is surprising to see how much creative spark can come out of this. Unfortunately, such creative spark is easily killed by a solely financially driven-budget cut exercise, especially if it's already the umpteenth such exercise.

5.1.10 The secret: Combine sensible budget cuts with instilling a creative constraints atmosphere

My strong belief is that minimizing costs, dealing carefully with agreed-on budgets and being prepared to take unexpected cuts to such budgets has to be or become part of the genes of every employee in a company, especially valid for those who are not producing "hardware" results like manufacturing, marketing, and selling products but who rather deliver what I would call "software," or in other words findings, discoveries, new ways, and simply new products, processes, and services. The latter group literally sits on a thinner branch when it comes to cutting budgets and, in my experience, R&D groups are often the hardest hit in such budget-cutting exercises.

Let me finish this section by that I strongly believe that sensible budget cuts, carefully planned and communicated, can be positive, whereas erratic behavior by top management announcing yet another budget cut because "times are difficult, exchange rates are against us, consumers lack lust" or similar excuses are extremely counterproductive to continued success in the marketplace, not only because it demotivates many or most of the company's employees (except maybe the CFO) and additionally gives a bad image to the company. Budget cuts have to be carefully planned, properly and constructively communicated, and creatively implemented to instill a positive, dynamic, and forward-looking spirit throughout the whole company. If outside consultants had a role to play, it would be exactly in this area, and to save the honor of some companies and some consultancy firms, this happens from time to time, not too often though. This brings me nicely to the topic of the next paragraph in this chapter, namely the general role of consultants and how they can and do influence the company and especially R&D.

Figure 5.2 shows an idealized illustration of how budget cuts (sensible and unsmart ones) and constraints can work together—or against each other for that matter. Please note that the system is flowing and the borderlines between the four quadrants are really for illustration purposes only.

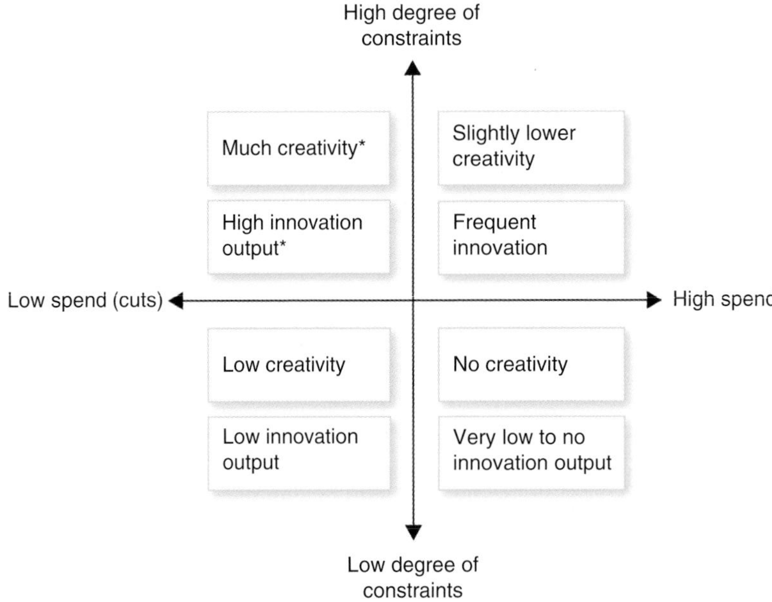

Figure 5.2 Budget cuts versus constraints – budget cuts and constraints.

5.2 R&D UNDER THE INFLUENCE AND GUIDANCE OF CONSULTANTS

Let me play a little word game here: if you take away the first three letters and the second "t" from the word consultants you end up with "sultans." Not to be taken too seriously, sultans typically are monarchs or kings, and when Mark Knoepfler and Dire Straits sang "We are the sultans of swing" back in 1978, they really meant that they are the kings of swing, whereas all other musicians just played the guitar, not too well either. The real meaning of *consult* from its Latin origin is rather seeing, looking or looking something up. A consultant should therefore be someone who sees and looks, sometimes—or often indeed—picks someone's (yours) brain and then comes back to you as the "king of new proposal" as to how your organization could be reorganized, newly structured, or simply with proposals regarding new areas of interest for the R&D community and how to best sell these to management.

5.2.1 Consultants sell you back your idea; What's wrong with this?

We typically like to poke fun at consultants by pretending that they come and listen to your ideas, repackage them, and then sell them back to you; there certainly are consultants that do this and you can see through this quickly, unless you are blinded by the beauty of the presentation of ideas. By the way, such an approach is often deliberately chosen by top management: the board feels the need to restructure a branch of the company, a department, or anything similar such as announcing a budget cut. Consequently, a consultant firm is invited to listen and look and pick their brain and is asked to come back with their proposals, which often are just a superbly packaged blueprint of the original request of the company's board. Once the results (the great package that is) are back in, the CEO and his (rarely her) colleagues come out with the announcement that consultant firm XYZ has discovered a wonderful and efficient solution to a problem that has befallen the company and can be resolved by this proposed way forward, and every employee is requested to give this his or her full attention and similar language.

Generally speaking, this is not a bad approach, although it typically costs a lot of money because consultants are not cheap; at least in my many years in the industry I have never come across a cheap consultant firm. On the other hand, expensive does not automatically mean that some of them might not be worth their money; some actually are. Let me discuss this in a bit more detail in the context of an R&D or technical group.

Like the movie title says rather dramatically, there are probably good, bad, and "ugly" consultants to be found. Good and bad consultants are rather straightforward to detect, whereas the ugly I mean are those who are not just bad and don't deliver anything sensible and constructive but become counterproductive and outright destructive. Luckily, there are not too many of those, but they do exist and they are often hidden behind a well-meaning façade. They promise to do the best for you and your group that apparently needs a consultant; in reality they work behind your back and possibly for your boss' boss. This can become a tricky situation because at first sight they look OK and at the end you might find your department being outsourced as a result. Tough luck some may say, I would rather call it *corporate deception*.

5.2.2 It's you or your boss who asked for help

Anyhow, let me get a bit closer to the topic of external consultants in R&D. There are almost always two possible starting scenarios:

- You feel that you need external expert help to redirect your field of activities or reorganize (to avoid the term *restructure*) your work group or department
- Your boss or his or her boss has these ideas and suggests to hire and finance a consultant for you to help you achieve the difficult tasks ahead faster and especially more efficiently.

There is a third scenario though when bosses themselves want to play consultants and not only instigate such changes but also throw themselves into the action. However, let me first stick to the external helpers and discuss and analyze how they may best be used irrespective of whether you were the initial requester for their help or your boss (or your boss' boss).

5.2.3 Consultants well used can be of real help

In general, there is nothing profoundly wrong with external consultants, even if sometimes or even often they are seen as threat by the target group of the consultants' intended actions.

One also has to realize that for a consultant working for your company is never meant to be a one-shot event; he or she wants to come back after a successful intervention, so it's in the consultant's interest to be as constructive and productive as possible. This is even more valid for consultants in the world of R&D. Their typical profile is not based on a business background—although this would often be desirable—but rather on a scientific and technical background. Ideally, consultants to R&D groups should be a hybrid between science and business. By having such a skill portfolio, they would be in an ideal position to help not only improve the dealings or structure of your group but also render it more understandable and credible for management at any level.

The general observation with regard to consultants and R&D groups is simple: make the best out of it or rather them; they can help and build and improve however, if not properly used they can destroy and pull down and even put blame on you so that they come out unscarred, ready to come back again and do their next thing. So the real question is: how can R&D properly use R&D specific consultants? The quick answer is simple: constructively appropriate and integrate their skills and know-how. Even if you have the feeling that you already know what they tell you. Part of such feeling may be justified, but then there is always this other perspective, this different and potentially novel point of view that one has not thought of, the surprising statement coming from an external expert that triggers a new direction of thinking and acting.

Unlike business consultants, R&D consultants are more of a trigger and less of a reorganizer, although the consequences of their interaction may lead to a new organization, a new structure, and potentially new people to be hired and others to be let go.

5.2.4 Being coached is everything

From my personal experience with consultants in R&D, I am convinced that the most efficient ways of consulting are not simple know-how download but active coaching and interacting with members of the company's R&D group in humble yet determined ways. This does not necessarily mean that external experts are never perceived to be arrogant, they often are. The main goal when working with consultants is to being helped to find better, faster, more efficient, and more innovative ways and results in one's work, which in R&D mainly consists of discovery, creativity, and innovation. If you as the member of an R&D team refuse to being coached by an external expert you are probably making an error: you should let it happen and not only make the best out of it in a "let-it-pass-by" kind of attitude but wholeheartedly accept that you may learn something new, and it almost never is a waste of time. It can be but, based on my personal experience, in probably more than 80 percent of the cases, R&D consultants are actually doing a good job, provided that they mainly stick to coaching and do not attempt to influence the organization and its structure. The exception would of course be that if they find real flaws in the organization, which are originating from wrong people organized in inappropriate structures the consultant has to flag it up. This then may lead to unpleasant situations and is probably one of the main reasons why consultants may have bad press that travels ahead of them.

Another important aspect that requires some discussion is the wrong approach of some, often many, members of the R&D who almost develop a kind of Stockholm-Syndrome when working with consultants: they are getting too close, too friendly, and too cozy and end up rather working for the consultant whereas the real solution is to work with them and use them for what they are there: their experience, their expertise, and probably their larger view of the industry in which the R&D group operates.

5.2.5 How to bring it to the consultant

Consultants come and go, but some come all the time and they do come back, again and again. There is a fine line between being welcome and overstaying the hospitality. Some consultants don't get this and they come back—or rather are being called back in again and again—simply because they are friendly with one of the higher echelons in the organization and have the hang and ease of proposing always new reasons why their consulting needs to continue. That may not be a bad thing; it can actually be really helpful but it may become counterproductive because the R&D group may get the feeling that they lose time. However, there is a subtle difference between rejecting a consultant because you believe that he or she has spent too much time with you and your group already and rejecting the consultant because you think you know it all. The latter is a feeling that I have come across many times with many colleagues in R&D in the past and could actually be the topic of a separate discussion and analysis, both of which I do not want to embark on here. Suffice it to say that whenever you have enough of a consultant because you feel that you have seen too much of him or her already, always ask yourself for your real motive. Be honest with yourself!

Anyhow, there comes the moment when you have really good and valid reasons to let a consultant go, honorably, profitably (for both sides), yet definitely, at least for quite some time to come. This is not always an easy feat because most consultants, good or bad, are typically well connected to your company, typically to the higher echelons and letting the consultants go can become a tricky and almost diplomatic undertaking. But it has to be done; it best is to be open and frank about the termination of their mandate. It's not a good idea to try to make it easier for the consultant by insinuating the promise of a follow-up project, unless you really mean it. Ideally, and that's probably something that you would do anyway, discuss such termination with your management and prepare for the final discussion that is planned with the consultant. Be straightforward and let no doubt come up that this is it and if you ever need the consultants again, you would let them know.

After all, consultants work for you and not the other way round, and because they are typically well paid, it should not be too tough for them to understand that the project is finished. I have seen consultants trying to pick up the ball immediately and try to meet with your management. Hence it is really important to prepare management well and have them on your side in such situations. The whole process sounds rather time consuming and inefficient, especially when one considers that the outcome of the collaboration was maybe weak, to say it diplomatically. A lot of effort for little effect but then one could never know this before. One might have obtained really important results from the consultants' contributions, obviously hoped for from the beginning. On the other hand, would the results be positive, there might not be a need or desire to terminate the collaboration. So, the unpleasant reality is that most effort of dealing with consultants, during or after the project that they work on for you,

▶ Redirect field of activities

▶ Pressure from upper management

▶ Upper management seeks personal glory

Figure 5.3 The 3 main reasons to hire consultants.

stems from unsuccessful collaborations. Tough luck after all, yet inefficient and another reason to change something in the way that R&D in the food industry works and operates.

Figure 5.3 shows in a simplified way the main reasons to hire and work with consultants. There may be many more detailed reasons, however from experience, these seem to be the three main ones.

5.3 R&D UNDER THE TUTELAGE AND GUIDANCE OF MARKETING AND OPERATIONS

I use the word *tutelage* on purpose, although it might be a bit strong. Others may call it *leadership* or simply *influence*. I want to show this strong and often unidirectional interaction between R&D and other functions in the food company such as marketing or operations. There are of course other players involved, such as finances, to name one more really important one in this context. Tutelage has several softer or slightly stronger meanings; I use *tutelage* as the state of being under a guardian or a tutor, or in other words, I suggest that R&D in the food industry is typically under a strong guiding, if not directing and ordering influence of other players in the company, especially marketing, operations, and the finance department.

I believe that I said this before: *marketing has* budgets, R&D *gets* budgets. There is a subtle difference between *has* and *gets*, not only from a linguistic point of view but also from a simple day-to-day business reality point of view. You may argue that to have you have to get it first and that is certainly true; however, there is a difference between *having* like in "I can pretty much justify every action that requires a specific budget because its always product related" and *getting* a budget to support the process of creativity, discovery, and innovation. The latter elements are quite naturally less tangible ones and therefore much more difficult to justify and defend. Therefore, it appears that R&D is always in this role of "may I kindly receive" when it comes to budgets because the more distant future is difficult to describe in exact numbers and almost existing results, which marketing can more easily claim for them.

5.3.1 Marketing has greater leverage

There is nothing fundamentally wrong with this; however, I have experienced many times that marketing claims a tutor role vis-à-vis R&D and lets R&D feel that they are the little brother (or sister) at best. The reader may think that this statement is the reflection of an inferiority complex and I gladly let you think this. On the other hand, I describe lived reality as a starting point not only to understand such reality but also to build it into any R&D strategy when it comes to dealing with marketing. By the way, the role of operations in this, what one could almost call "power game," is a different one and I shall discuss and analyze this a little bit later in this section. Like in most potentially conflicting situations in life, be it business or private, it is important to know one's starting level of interaction, negotiation, and sustainable and satisfactory solution. I avoid the often used term *win–win* purposely because I do not entirely believe that win–win can actually be achieved with all its necessary compromises and consequences. It may happen sometimes; however, most of the time, it is more of a nicely sounding lip service rather than hard and simple reality.

I do realize that compromise is seen as something desirable and necessary to achieve in negotiations in all situations of life. And compromise is probably also necessary to achieve otherwise everything would come to a standstill and nothing moves forward. On the other hand, however, compromise also means that neither side had a good and convincing enough proposal that was so obvious in its outline, clarity, and promise of success and that would have swayed one party completely over to the other side and no compromise was necessary, actually only little discussion was necessary. The need for compromise always means that neither proposal was really good and wooed the other party. The next best solution is then obviously to reach a compromise and both parties give in a bit and can recognize bits and pieces of their original proposal in the final, agreed-on version of the proposal for whatever it may have been: a project, the acquisition of equipment, the hiring of new personnel, the increase (more recently rather decrease) of research budgets, and similar situations. The results are most often euphemistically called win–win and comments such "we are very happy with the resulting compromise and have reached a full win-win for both sides" are typical.

5.3.2 Marketing gives orders; marketing does not make compromises

Again, I want to emphasize that compromises are a fact of life but are not really a good solution to resolve diverging opinions. Marketing people have understood this and that's probably one of the main reasons why in negotiations with marketing there is little compromising to be found, if at all. That is also the reason why R&D people almost always have the impression that they have no say in the dealings with representatives from their marketing department: they tell them what they want—better require, desire, need—and R&D or the technical branch of the company has to deliver. That is also the origin of the well-known "Marketing—Technical Ping-Pong" game, so well and frequently played in many corporations.

I have extensively written about this for the case of interactions between marketing and the technical packaging group in *The Food Industry Innovations School* (Traitler 2015, p. 193):

> *Playing ping pong*
>
> *Maybe it's time for a little personal story here. When I was in charge of packaging in the company's largest market organization, we had a kind of thing going on between technical packaging, as it was called, and marketing. Every single step of packaging development was not only firmly requested by the marketing person or team, but was closely followed and inspected; really, every*

single step and then some. It typically turned into a Ping-Pong kind of competition. The packaging team (PT) proposed a mock-up and sent it to the marketing team (MT). The MT received it and, first thing, sat on it, almost like a Greek-coffee/Turkish-coffee situation: take it off the flame and let the grounds settle.

The MT came then back to the PT and asked for the next iteration, such as smaller, less material, different material, modified mock-up, easier opening, etc. You name it, you got it. This typically went through several iterations until someone thought it was good—or maybe had just had enough of the game—and accepted the solution.

Everyone realized that this was not a good situation, but no one really wanted to let go of the power that comes with having the last word. Out of curiosity we did a kind of time progression analysis, as I called it, simply trying to measure and assess the time spent in the individual steps of:

The PT making draft prototype → sending material to the MT → the MT letting it settle → the MT taking first round decision → the MT refusing first prototype and deciding to send it back → the PT receiving refused first-draft prototype → the PT making second prototype → the PT sending it to the MT → the MT receiving second prototype and letting it settle → the MT taking second-round decision, refusing and sending back."

The conclusion of that discussion is simple: marketing gives the orders, and technical (including R&D) has to deliver in such a way that marketing is satisfied. This can go through several lengthy, painful, and probably costly iterations and clearly shows what I have suggested: "marketing gives orders, marketing does not make compromises":

Let me again bring this back to the theme of this chapter and especially of this section: inefficiencies of R&D. The suggestion that R&D and technical just has to deliver and has no say in these matters is of course not only simplistic but also partly wrong or rather wrongly understood. R&D is not the proverbial sheep that follows the herd and delivers its wool, milk, or meat. No, R&D can be, and in many cases often is, the tip of the spear that is leading toward the future successes of the company. This is, however not achievable in "suffering follower mode" and change is needed, real change in attitude, composition, and project portfolio of the R&D department of food corporations, small or large. Much of this will be analyzed and discussed in the following chapters, especially in Chapter 10.

5.3.3 Operations act like a strict father

In a previous paragraph, I suggested that the interactions between R&D and the operations (the manufacturing departments) differ from those with marketing. There is more psychology behind this. Let me discuss what I mean in more detail. Maybe a short definition of what I mean by *operations* should be in order. In most food companies, like in many other industries, operations or often also simply called *manufacturing* is that part of the company that is in charge of the technical assets and their usage to manufacture products. The branch of operations in most cases also includes logistics and supply chain, both in-bound as well as out-bound supply chains. In other words, procurement of raw and packaging materials, transport to and from the manufacturing sites—the factories or plants—is under the responsibility of operations.

The core business of operations is, however, the manufacturing part (i.e., making products on the assets, the machinery that operations has at their disposal). These assets include the brick-and-mortar factory as well as all technical equipment inside and outside the factory buildings. Thus far I have not mentioned people.

Operations are typically managed and run by technical experts, mostly engineers of all sorts, such as mechanical, electrical, and probably most of all chemical engineers, at least such is the mix in the food industry. Typically, these technical experts are supported by experts in logistics in areas such as transportation and storage as well as procurement. There may be additional expert areas that I have forgotten in this short description. This is of no real importance because the operations department is fully integrated and embedded in the company at large and is supposed to work with all other branches in the company, especially R&D, including quality and safety, as well as marketing and sales. Operations is also closely supported by the legal department as well as communications experts because in case of quality problems with products, recalls, or other problems such as health- and safety-related ones, one person in the chain of command of operations is ultimately responsible for any such incidents. Typically, this is the highest-ranking technical officer of the geographical entity (the market) in which such an incident happened.

5.3.4 A bit of humor

I explain all this because it partly explains why operations in most food companies are a tightly knit community that prides itself to be able to solve all problems with the help of their own operations-based experts and won't let outsiders, especially scientists, come too close to their world. This is the point at which the psychology part, which I have mentioned before begins. There has been an ongoing state of competition between scientists and engineers, which is founded in the age-old history of these disciplines. Let me support this by a few more or less serious, more or less funny quotes from both sides:

> *In science if you know what you are doing you should not be doing it.*
> *In engineering if you do not know what you are doing you should not be doing it.*
> *Of course, you seldom, if ever, see either pure state.*
>
> Richard Hemming, 1975

> *The scientist describes what is; the engineer creates what never was."*
>
> Theodore von Kármán, 1980

> *Engineers think that equations approximate the real world.*
> *Scientists think that the real world approximates equations.*
>
> Unknown

Let me close this short discourse on the subtle differences between scientists and engineers with the simple remark that for a company that is operating in manufacturing food both disciplines are absolutely necessary and required for the present (engineers) as well as future success (scientists) of the company. Sadly enough, I have still observed a lot of castle-building, by both sides I have to admit, where it becomes extremely difficult for engineers from operations to enter into the more scientific world of R&D and even more difficult, close to impossible, for a scientist to enter the closely knit world of operations.

5.3.5 Here's one example

Let me illustrate the latter with an example: factory managers (or factory directors) are typically or almost exclusively recruited from the world of engineers that were originally hired into the department of operations. In all my years, I have never seen a scientist become factory manager. Let me correct that statement: there was one exception, however, I am pretty sure that this was the only one given the large number of factories in my former company, which amounts to approximately half a thousand factories worldwide. The excuse was always (without exception this time): scientists are not and will not be capable of understanding the difficult operational situations to be found in this "minefield" of operations, in this case, manufacturing. I have personally worked in operations, although on the corporate side, so I know factories mostly from my numerous visits and having spent weeks in a row for training purposes in especially chocolate as well as coffee factories of my former company. So, I am not exactly an expert on what it really takes to become a successful factory manager. I do, however, know, what it takes to become a successful scientist in a food company's R&D group: many years of university training, typically including a PhD, and many more years, probably at least 3 to 5 of internal company training to really deliver not only great ideas but really innovate and realize ideas in the form of products, processes or services for the company.

It would be in the strongest interest for any company to foster exchange of their experts in different disciplines, especially also exchange of scientists into typical operations roles as well as engineers working in typical R&D development projects. Alas, it hardly ever happens, at least in my experience, and it's mostly got to do with this aversion of representatives from operations to accept scientists into their ranks. I say this in such a unidirectional fashion because I have seen the opposite time and again: a large number, if not majority, of R&D directors, especially of typical product-related R&D centers come from the engineering, the operations side. This is yet another reason why present day R&D groups in the food industry are inefficient: their representatives lack opportunities to be exposed to the operations and manufacturing side of the company, and consequently they miss great personal development opportunities, which would be necessary to strengthen and improve the efficiency of the R&D group.

5.3.6 Let's be respectful with each other

I have sat in many meetings in which rather "demeaning statements" were made by representatives of the manufacturing teams targeted toward the R&D community, statements that were not helpful in cementing the unity of the different branches in a company or in improving the efficiency of all its branches. This efficiency improvement will only be possible if all sides begin to understand that there is no need for castle-building, for whatever reason (Feelings of inferiority? Simple fear? Feelings of superiority?) and that the only successful way forward is that, of what I would call "differentiating complementarity." What I mean by that is simply giving the space to each group in the company to make their marks, shine, and rise at the same time because of the complementary nature of all work, it fosters exchange between the different groups. This topic will be discussed in more detail in the following chapters of this book, especially in Chapter 10.

5.4 EVOLUTIONARY CHANGE IN A TYPICAL FOOD R&D ORGANIZATION

This last section of this chapter is really an introduction into the following three chapters, which will discuss and analyze in much detail the needs and suggestions for change from the consumer, the university as well as the industry perspectives, respectively. But what the entire discussion in this chapter suggests strongly is the fact that present-day R&D organizations, for the number of reasons that were elucidated, are inefficient, largely underperforming and require modifications, if not to say radical change. I do realize that such a statement will provoke quite a few defensive reactions and especially explanations why this is not the case, such as R&D groups, after all, supply their parent companies with sufficient innovation and are responsive to every need of the company. And such statements are certainly partly true and correct but in their totality they don't hold water. The unpleasant truth is that today's R&D organizations in many food companies are inefficient and underperforming.

5.4.1 R&D is not alone in mediocrity

These statements target the R&D group in particular but that is not to say that every other branch of typical food companies is perfect and performing at its best; this would be far from the truth but this book is after all about R&D and how it could be improved and that's the topic I intend to stick to here. Unfortunately, the parts that are efficient and perform are always cited to explain and justify the performance level of the entire organization. It's a bit the same like an alcoholic; there are sober times but overall, the person still drinks too much and depends on drinking. Until such time that an alcoholic does not admit that he or she is actually a dependent alcoholic nothing will change. Only if self-recognition sets in, things will change. It's the same with organizations of any kind: until such time that there is recognition of inefficiencies, no corrective action will begin.

Nobody speaks about revolution here (at least not until Chapter 10), so there is no need for fear and rejection of the inevitable: evolutionary, subtle, and smart change is required, but required it is. For those who are involved in the organization it's not only a tough recognition but it's especially a tough call to do "something" to get it right. This book, should help to explain and define what this something really is. However, it's not only about the something, it's also about the "how" and the "when" and the "how much" and "who" and possibly a few more of these, which I shall begin to discuss and analyze in this last section of this chapter.

5.4.2 Let's change, gradually!

Let me briefly discuss what I mean by *evolution*, by *evolutionary* in this context. The *Oxford Dictionary* gives a nice and fitting general definition of the word *evolution*: "the gradual development of something." I do like this definition because it carries all the necessary elements:

- Gradual
- Development
- Something

Here we go, it's all there; the word *gradual* suggests that any development of something will only happen over time, gradually, well thought of, much discussed, and finally agreed-on by most. I do realize that I embellished a bit here but honestly, that's how I feel, when I hear the word *gradual*. It suggests this phase of fact-finding, analysis, discussions, and defining the new elements, and after, deliberation and agreement, gradually putting them into practice. Bingo! Sounds nice, doesn't it? Well, I have lived through quite a few evolutionary steps of organizational improvement attempts, and they were far from easy, agreeable, and fast; but at least they always were an attempt to make things better. If the only remaining positive element in such attempts was the one of recognition that things were suboptimal and needed to be changed, this was already worth the exercise.

5.4.3 Watch out for support and best timing

Any proposal for change needs to be sold to some decision maker who ultimately will not only agree but will also put necessary resources behind such change, be it people or money. Typically, such agreements will be given by top management, and it would therefore be a good approach for anyone in the organization who wants to instigate change, change for improvement that is, to include many hierarchical levels in the organization and take their points of view and especially their doubts and fears into consideration.

Another element in any change process is the element of timing—I mean really good timing—if the change process is expected to turn out successful. But what is good timing, timing that is the most promising one? The answer is not a simple one as it much depends on the situation. Let me turn to the elements listed in Figure 5.1. If the reason for change lies for instance in the perceived flaw of results coming too late, then optimal timing for change would be now. The remedy mentioned in the figure is increase speed. That's easier said than done and requires smart actions such as focusing on fewer projects, in-sourcing additional expertise or, if possible and appropriate, hire more experts. The latter in today's environment is probably rather difficult to do and should really be the last resort in case everything else fails. Also the timing of *now* would be difficult to realize; however, if turning to the other actions just mentioned, execution of these could almost happen immediately.

5.4.4 Cyclical versus anti-cyclical

The perceived flaw elements in Figure 5.1 can partly be remedied by proactive action taken by members of the R&D group and partly depend on goodwill and actions being initiated by management. The latter is no excuse, but it has to be said, taking the element of "cyclical investment approach" as one example, a real change toward an anti-cyclical approach has to clearly come from management, which is against the odds that the financial analysts would have the guts to dare such a new direction. Let me expand on this thought a bit more. I have already briefly suggested that companies may want to go to an investment approach—into their R&D that is—in an anti-cyclical way: when business is doing well and earns a lot of money, tone down the R&D investments into their usual product areas and invest into potentially totally new fields and product areas with great promise for the future. Such an approach could help prepare companies to better ride out periods of weakness in the marketplace because the innovation portfolio is such that it can readily be deployed whenever needed.

From my personal experience, I have seen the opposite happen: the more a particular business is successful and doing well, the more management is willing to put more resources—money and people—behind. This is totally understandable for new businesses that are in a building and growing phase but makes absolutely no sense for mature businesses, several decades old already. But life is so much easier and one is definitely so much on the safer side when doing the "obvious," namely support success even more because this, by perceived definition will bring more success. All that it really brings is most probably the mature and decades' old business to continue to be fairly successful but not more. Let me say it in numbers: When the business, on a scale from 0 to 100 is at a success level of say 60, by investing the equivalent of 10 (resource wise) it might just keep its success level of 60 or maybe with lots of luck, climb a pint or two.

5.4.5 From 10 make 1 or make 10: Which do you prefer?

However, if one would invest the same equivalent of 10 to completely new areas of research and development, still perfectly fitting into the overall framework of product and service activity of the company, one might start at close to zero but the investment could bring this new area easily up to 5 if not the entire 10, or with some luck and strong determination, why not even higher.

What I want to demonstrate here that R&D investments are more wisely spent into the novel than just throwing it after the known. Unfortunately, the latter is what mostly happens today, which also, at least partly, explains the unwillingness of food companies, smaller or larger, to invest into startup companies or create their own incubators.

As the present section of this chapter is all about evolutionary change I am perfectly aware that suggesting such a strategy change to the traditional R&D investment approach is far from evolutionary and is therefore maybe not realistic to achieve in standard business situations that a more or less successful food company is finding itself. Real substantial change is easier to achieve in extraordinary situations and times such as companies in distress or at least perceivable real difficulties. Not that I would wish such situations for any organization, but it is true and historical examples can support this that in times of turmoil, evolutionary and slow and cozy change mechanism can dramatically and fast change to the more revolutionary of disruptive change, change that will be discussed in more detail in the following chapters.

5.4.6 Let us team up!

Let me, however, pick up a few more elements listed in Figure 5.1, such as "no good understanding of business needs." That's a simple one and actually easy to fix, provided that the different branches of the company, for instance marketing and R&D or procurement and R&D or some other "pairs" really talk to each other, respectfully and with the understanding to grow together. That, in my experience, is unfortunately not always the case but can be remedied so easily and so quickly. I have only recently met with a small group of R&D representatives who have expressed the desire to learn more about and better understand the businesses for which they work but at the same time expressed the difficulty to find the right, and most of all willing speaking partner within their own company. Such is the reality, at least in larger organizations. This is one of the major differences to smaller food companies where there is a much closer physical proximity of the different branches of the company a daily reality.

5.4.7 Change comes easy

The real evolutionary change in large organizations is simple: management to foster and encourage the increase of the number of nods, of connecting dots between the various business groups and R&D and this not in a unidirectional, ordering way (marketing to R&D, procurement to R&D, finances to R&D, etc.), but in real partnership ways and by this ensuring that both sides gain the maximum of understanding of the other side and through this quite naturally involve R&D increasingly in business decisions. A similar remedy, namely increase interaction and integration—increased involvement in the business—is valid for another perceived flaw from the list in Figure 5.1, R&D is perceived to be too far detached from business.

This latter perceived flaw, together with the perceived flaw of R&D having no good enough understanding of the business can be remedied easily and quickly without any extra costs nor other hardships. It requires goodwill and desire and willingness to work together for a common goal, namely to take best advantage of the exciting results that a great R&D group can bring to the company in the simplest and fastest ways and for the ultimate financial benefits of the company. That's not too much asked for or is it? Most of the time it's big egos that stand in between such perceptions and their remedies. It's about time to get rid of such egos or put them in their rightful place.

5.5 SUMMARY AND MAJOR LEARNING

- This chapter discussed and analyzed the all-important topic of the need for a new approach to R&D in the food industry. It discussed and analyzed the critical topics of inefficiencies of R&D; the role of consultants for the company at large, especially for R&D; and the overpowering roles of marketing and operations and evolutionary changes suggested for the R&D group.

- When it comes to R&D, top management seems to be more concerned with the costs of R&D rather than its output. Constraints, which are generally seen as healthy and creativity fostering are often confounded with undifferentiated cost reductions.
- Mistrust toward R&D is often an overwhelming sentiment for top management and drives many of management's decisions. And this despite the fact that it is always members of management who hire personnel, the same personnel that is then met with mistrust.
- It is suggested that R&D in the food industry, especially in larger food companies, is systemically inefficient and top management likes to point fingers to the R&D group frequently, possibly to distract from other inefficiencies in other parts of the company, but why not themselves?
- One major and real reason for the inefficiency of a food industry R&D group is its desire to do too much, too many topics, too many projects, too much of "everything.".
- The influence and role of size of the company and especially its R&D group was briefly discussed, and it was suggested that smaller entities are typically less inefficient because fewer people have to necessarily become more versatile to solve problems and create innovation. However, size is only a minor element in this discussion of inefficiency.
- The important question of "what's wrong with R&D" was discussed in Figure 5.1 as well as in a later segment on evolutionary change. A list of possible perceived flaws of the R&D group was compared to a list of possible remedies. There are three major groups of flaws and remedies: first the group of flaws that can be remedied fairly quickly and which is within the sphere of influence of the R&D group itself. Secondly, there are flaws that are totally outside the direct influence of R&D and are either system immanent (e.g., growth of the group through acquisitions) or can only be remedied by willful action of top management. Lastly, there are a few perceived flaws that need a dramatic and disruptive approach to correct such as introducing an anti-cyclical investment approach.
- The desire, sometimes almost fashion, for budget cuts was discussed and it was suggested that undifferentiated budget cuts often seem to be the ones that happen more often than those which take real needs and requirements of individual groups into consideration. Such undifferentiated cuts were compared to shortening a text to half its length by striking every second word.
- It was suggested that by combining sensible and symmetric budget cuts with instilling of a creative constraints driven atmosphere the overall creativity of the R&D group can most likely be lifted in positive ways.
- The role and guidance of consultants was discussed and analyzed and three major reasons for hiring consultants, also in the world of R&D are: redirect the field of activity, pressure from upper management, and upper management seeking personal glory. Although the third reason may sound cynical, it can often be observed.
- The equally important question how to terminate the mandate of a consultant was also briefly discussed, and it was suggested that honesty and a straightforward approach, without burning bridges for good was the most successful one, and the least prone to criticism by upper management.
- The strong and influencing role of marketing and operations was discussed as an important example how R&D is often almost put under the tutelage of other groups. Marketing,

especially operations, are tightly knit communities in every food company, and it is not easy for R&D to make its marks and play the role of an equal partner.
- The question of cyclical versus anti-cyclical investments into R&D was briefly introduced and discussed.
- To change and improve the inefficiencies of especially larger R&D organizations it was suggested advisable to start with evolutionary change by correcting perceived flaws such as low perceived output, late results, being too far detached from business, or not having a good understanding of business needs. All perceived flaws can be corrected by the R&D community fairly easily and rather expediently.
- Finally, it was suggested that change comes easy, and should be embraced by the R&D community without hesitation. Such change can be helped by multiple interactions between many members of many branches of the company and R&D is a great environment to start such change.

REFERENCES

Traitler, H 2015, *Food industry innovation school: How to drive innovation through complex organizations*, Wiley-Blackwell, Hoboken, NJ.

Traitler, H, Coleman, B, Hofmann, K 2014, *Food industry design, technology and innovation*, Wiley-Blackwell, Hoboken, NJ.

6 Consumer perspectives for change to R&D in the food industry

We used to think that the enterprise was the hardest customer to satisfy, but we were wrong. It turns out, consumers are harder than the enterprise because the consumer will not give you a second chance.

Eric Schmidt

6.1 THE FAST MOVING CONSUMER GOODS INDUSTRY (FMCGI)

A lot is said about the fast-moving consumer goods industry (FMCGI), such that it is actually slow moving; that it carries and suggests consumer goods that consumers already know; that nothing really comes from the industry other than "pushing products down the throats of consumers"; that industrial products, especially when it comes to food and beverages, are not good anyway; consumers pay way too much (actually these days probably way too little); and that the products are not healthy, potentially not really safe, make you fat, and so on.

This litany is an undifferentiated view of the situation and especially the situation of the food industry as the main representative of the FMCGI. We could also simply call this industry the *packaged goods* industry. On the other hand, everything seems to come packaged these days, even brides and grooms in their gown and tuxedo and nobody will really call them parts of the packaged goods, although given the divorce rate in many civilizations one could believe so. So, bride and groom, who are consumers too, are part of a fast-moving industry, at least to some degree: the fast-moving consumer or packaged goods industry.

Food Industry R&D: A New Approach, First Edition. Helmut Traitler, Birgit Coleman and Adam Burbidge.

A short definition for fast-moving consumer goods or consumer packaged goods follows:

These are products that are sold quickly and at relatively low cost. Examples include non-durable goods such as soft drinks, toiletries, over-the-counter drugs, toys, processed foods and many other consumables. In contrast, durable goods or major appliances such as kitchen appliances are generally replaced over a period of several years.

6.1.1 Fast, furious, and cheap!

The emphasis here appears to lie on two elements: costs (or rather price) and speed ("quick sale"). The third, important element is only defined by exclusion, namely frequency of sales. It is this frequency of sales and the share of sales of products of fairly low value that really defines the FMCGI. You can have quick sales elsewhere; at the height of the financial and especially real estate crisis, when thousands of homes went into foreclosure, such foreclosed homes were sold faster to investors for all kinds of reasons, not least of all pure speculation, then one could imagine. By the speed-alone definition this type of product "foreclosed homes" would have qualified for the membership in the FMCGI. Even by the frequency definition, this might have passed for an FMCGI product. And, in relative terms, the costs were low too, at least compared to what they were before the crisis hit, however, in absolute terms they were (and are) of course much, much higher than a fairly low out-of-pockets spending for a confections bar, a cup of coffee, or a prepared soup. The latter, together with all their siblings in the food industry such as ice creams, milk shakes, nutritional beverage mixes, mineral waters, infant formulae, and many more, are the real heroes, the real members that make up this vast industry of packaged goods.

A typical large food company or other packaged goods company that is doing business worldwide probably sells several hundred million products every day, some probably sell even in the billion units per day range! The average sales price of the packaged good is probably anywhere between 50 and 100 cents, obviously depending on the type of product sold. These are really staggering numbers and give you an impression of the size of this industry, at least as far as the frequency of interactions with consumers is concerned. This is actually one of the great features of the FMCGI, the almost endless, numerous opportunities to interact with consumers on a daily, even on an hourly basis through sales as well as simple use of products.

This is one of the reasons why the "packaged" part—the packaging—is so important, if not outright crucial in the context of packaged goods, and much has been said and written about packaging being the last or rather the first interface to the consumer. I wrote several chapters about the role of design and packaging design in the food industry, and I can only suggest that you consult these chapters in case you would want to learn more about this topic (Traitler, Coleman, & Hoffmann 2014).

Let me just say this here: the role of packaging in most of today's consumer goods companies, and especially food companies, is highly underestimated and packaging is still and increasingly again seen as a cost factor that should be reduced and kept to an absolute minimum. This totally neglects the great opportunities that come with modern, efficient, and professional packaging design and can enormously support a product and its success in the marketplace. Packaging,

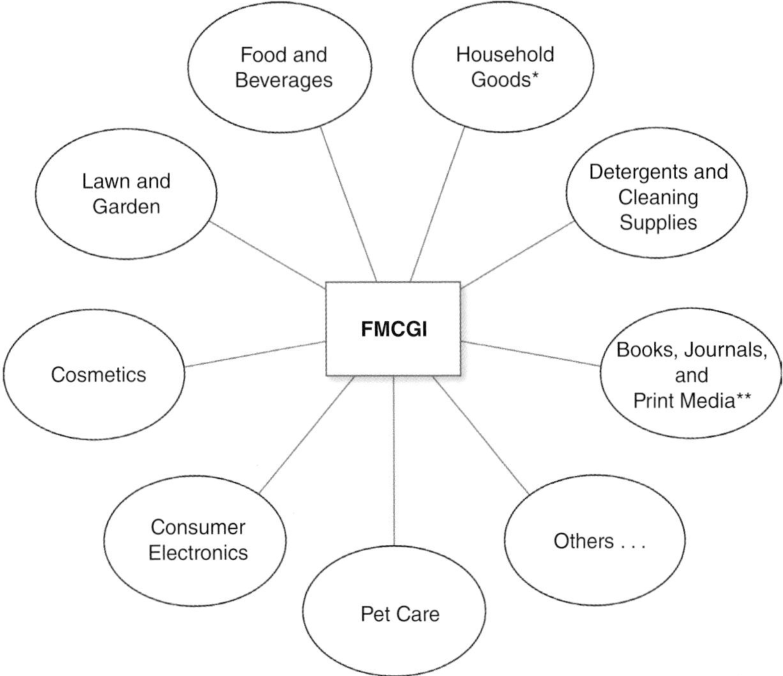

Figure 6.1 These are the industries of which the fast moving consumer goods industry (FMCGI) is typically composed.

especially in the R&D community, is almost seen as ancillary and is often considered and realized far too late in the research and development process.

Figure 6.1 depicts the typical family members of the FMCGI. Normally, print media and all related are not seen in this group, however, they seem to meet all the criteria.

6.1.2 What consumers really want? The million dollar question, the billion dollar answer!

But this is not a chapter about packaging; this is an open discussion and an attempt to analyze the consumer perspective or perspectives with relevance to expected, needed change of the R&D approach in the food industry. Although there is no single and unique consumer perspective, there are multiple facets of such consumer perspective and these are based on perceptions, hopes, some hearsay, and probably a lot of misinformation or incomplete information.

One of the driving forces for any industry and especially the food industry is "cost efficiency." The quest for cost efficiency seems to be at the forefront of the debate in the entire industry and is largely driven from two sides; on the one hand, there is the group of financial analysts who valuates the company by its cost structure and success in the margin race.

On the other hand, consumers increasingly want to receive their food and beverage products at increasingly lower prices. The latter is certainly also a consequence of the retailers' promise to the consumers that they will be able to purchase their food and beverages at the lowest possible price in their store. I do realize that this is called *competition* and should be good for the consumer, at least for his or her wallet. But is it good for the consumer when it comes to health and well-being based on good tasting, safe, nutritious and yes, affordable food? And these days there are many more requirements that modern consumers of food and beverage products ask from the food industry. Ideally, all food should be fresh and not industrially produced; and if it was industrial food, it should at least be as natural as possible, not contain any flavors or other ingredients that make for a rather lengthy ingredients list and that turns the consumer rather away instead of attracting him or her.

6.1.3 Food should be all natural it should be all this...

Food should be all natural, it should be organically grown, and in some areas of the world it's called *bio* (I do realize that there is a more subtle difference than just geography). For some consumers it should not contain lactose, for others it should be gluten free, yet for others it should not contain any meat, or at least not pork meat, again for others there are religious motives and drivers such as halal or kosher food. Other consumer groups follow one or the other of the hundreds of diets that you can find today. Yet others have allergies against proteins of all kinds such as milk or peanuts. I could probably continue with a long list of more consumer requirements and resulting expectations when it comes to *their* food and beverage habits.

It is difficult to find one's direction in all this. Especially because the food industry and its R&D arm seems to always chase after new trends, new findings, newly identified culprits for all kinds of health-related issues and finger-pointing has become a favorite pastime for many inside and outside the food industry. This may sound like an excuse to the reader and I have to admit that it partly actually is. It is truly difficult to find the right direction in this trends and fads jungle and give sensible and good things to the consumers and be able to do this consistently and always in high quality and at affordable prices.

And because it is so difficult many in the industry find excuses to either do nothing and hoping that the "fad will pass" or jump to hyperventilation and almost automatically select the wrong direction, mainly based on what competencies and know-how are already available in the company rather than to perform a serious analysis as to what really might come next and whether the company and especially its R&D organization are actually prepared for the new challenges. It reminds me of the drunkard who searches his lost car key under the lamppost because there is light and not where he has really lost the key. Unfortunately, this has a lot to do with the quest for cost efficiency and not so much with restricting resources and thereby becoming more creative and innovative. It is more complex than that and needs a few serious reflections.

6.1.4 Food companies don't like risks; they "wait them away"

One of the reasons why this happens the way it most often happens is founded in the well-known risk averseness of the food industry. This is not an "on-dit," this is a simple, yet sad fact.

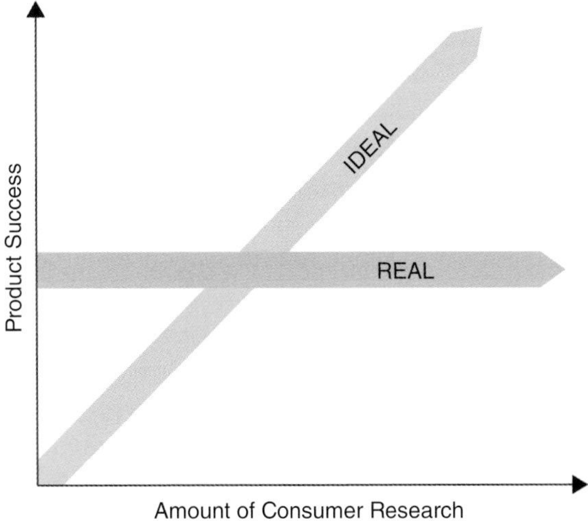

Figure 6.2 Product success versus amount of consumer research behind product development.

I have experienced such climate of risk aversion throughout almost all my years during which I have worked in the food industry. There were a few exceptions, and they had mostly to do with new, fairly young (in service years) and refreshingly inexperienced leaders coming on board and leading us against all odds to successful results, taking high risks all along the way to success or to failure. The real fun part here is that one or two real great successes by far outweigh the relatively large number of failures. The other unexpected observation that I could make is the following: the greater the success of a new product idea and development the farther it seemed to be detached from initial consumer insight and market research. That is not to say that consumer insight is not a good basis for new product development, but it does simply not make for large steps forward but rather for baby steps of improvement or what the industry calls *renovation* of product and thereby nicely brushes over the fact that it wasn't more daring in its quest for new products, processes or services.

Figure 6.2 illustrates the ideal and real correlations between the success of a new product in the marketplace and the amount of consumer research that went into the new product development.

I would suggest the existence of a negative correlation between size and importance of success of new products and the amount of consumer research that was available at the onset of any underlying project. Those who know me may say: yes, this is a guy who spent much of his time in R&D and loves his technologies too much and really wants to neglect the consumer. And I say, yes, they are right, but I have good reasons to at least neglect the consumers' wishes to some degree. If the industry only followed consumer insight and did straightforward product development along these lines, it would hardly need an R&D division that possibly costs them exactly the 50 percent too much. R&D is only really worth its money when it is used to excel and strive for the impossible, which normally does not manifest itself in the consumers' minds.

6.1.5 Lean and efficient: Don't you get it?

This is of course in stark contrast to the mainstream direction of any food company, or any company for that matter, namely the drive toward lean and efficient organizations. These are two of the most popular buzzwords that the industry uses over and over again: lean and efficient. Cynically, yet truthfully this can be interpreted as: as few people as just necessary (where the "necessary" is determined mostly by financial considerations) and as much personal, individual commitment as humanly possible. In principle, there is nothing fundamentally wrong with this approach, as long as it respects human dignity and the fact that every person can only be "used up" so much. However, if this fine balance of what is acceptable and what is beyond possible is clearly not respected, the talk about lean and efficient remains just what it really is: talk with no real substance behind.

This is not to say that lean and efficient organizations are not desirable or necessary, they are and this is especially true for an R&D organization in a food company. One could add further attributes such as smart R&D or oriented toward delivering results, ideally on time and on budget or even before and below. And it lies in the nature of the FMCGI that its turnover is fast, happens frequently, and renewal of its product lines and underlying and supporting processes happens constantly. This is the real difficulty to combine the need for speed of any FMCGI company with the need for depth and ample and sufficient time of every R&D organization. Disputes, misunderstandings, and frustrations are almost built in the system and therefore should rather be expected and should not happen to the surprise of anyone in the company. And yet, it so does; company executives always seem to be surprised that R&D, their smart, lean, and efficient R&D organization did not deliver as they had them expected to do. As always, the burden of fault is not entirely lying on the company's management side but is equally shared by the R&D organization. More often than not have I experienced the R&D group being elusive and defensive as far as timelines of delivery and costs of delivery are concerned and thereby trying the easy way out of an otherwise shared responsibility why projects have not succeeded as expected or why budgets were surpassed or other explanations, which I would not call excuses but rather attempts to get management to understand the situation.

6.1.6 Mutual understanding is not everything; it's the only thing

I expand on this point because it is an extremely important one in the cohabitation of R&D and the business side of the company. It's actually the real crux of the entire matter why the R&D organization is often belittled, denigrated, and bad mouthed by the company's management. This is a broad-brushed painting of the situation, but I have experience such comments as the R&D organization has no understanding of the business, does not understand the real need for speed, and especially almost always has too many people working on problems that were not requested by the business. Again, such comments are not totally unfounded but paint a too undifferentiated picture of the situation. There is, of course, a rather easy way out of this impasse: mutual respect and a real desire to understand the other party's position and needs.

When looking toward other FMCGI companies such as home products including home and garden, cosmetics, or even household electronics, it appears that the situation of the R&D group is not different from the one in food companies: financial pressures on companies to deliver timely and successfully—meaning with good profit margins—put pressures of all kinds on the R&D group, often asking for the impossible and tempting the R&D group to promise the impossible. And the temptation is great and members of the R&D group quite easily fall into the trap of overpromising. This chapter, however deals with the consumer perspectives and how they may influence or encourage changes in the R&D organization of a food company. So, how do consumers fit into this equation of a lean and efficient company organization, including its R&D group? If food and beverage products are perceived as commodities with everyday low prices and a "stinginess is cool" attitude by consumers, then it cannot be a surprise that any food company, irrespective of its size *must* strive toward a lean and efficient organization and that unfortunately includes the R&D group, even if this is seen as totally counterproductive and against the nature of R&D.

6.1.7 Here are some ways out

How then can R&D step out of the train of thought of reducing costs and resources at any rate and still, with increasingly limited resources deliver what is expected: innovative new products, processes, and consumer experiences? Well, there are several ways. These are not recipes to solve the riddle of how to make more with less, but they are hints as to how to rather simply and straightforwardly maximize the output of the R&D organization:

- Turn to fewer and initially already more promising projects. In other words, bet on success and focus!
- Say no to the business more often but give them exciting results so that your "no" can be accepted more easily and is no turn-off for management. In other words, show them what you got and say no to seemingly unreasonable requests.
- Anticipate requests for projects and simplify and streamline the project work. In other words, use available resources more efficiently by deploying them only to projects were they have "single input with double impact."
- Optimize the own R&D resources by increasingly turning toward exterior support by experts outside the company and its own R&D organization.
- Ultimately, accept constraint for what it really is: a source of inspiration and support system for creativity out of which typically grows great innovation. In other words, share knowledge and competence in more open ways in your organization and simplify processes (typically red tape), if necessary "kill" them.

Figure 6.3 illustrates the overall approach how to reach more with less.

6.2 THE CONSUMER IN THE CENTER

"The consumer is king" (or queen for that matter), is something that you must have heard time and again and it has truly been said and written again and again. But what is really meant by this? The consumer dictates, reins, governs, and ultimately decides on the fate of a product? And, by the way, who is the consumer? Is there such a person like "the consumer"? Personally

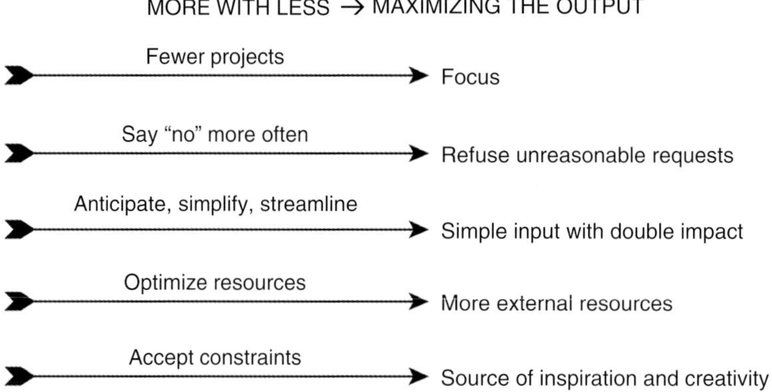

Figure 6.3 Maximizing the output.

I don't think there is; there is no one person that represents the consumer. I would admit that one can probably find a group of people, fairly homogenous in their habits and preferences, conditioned by a certain type of environment, including advertisement and trends, certainly personal experiences and hopes and yes, disappointments and negative experiences. Modern-day pollsters are efficient in discovering and determining such groups, especially in the case of politics and elections. A similar approach is used to determine and define consumer groups and asking them for what they want (outspoken needs) but also what they do not clearly want yet (unmet needs). The latter are of course more uncertain and risky and therefore by default much more difficult to justify to pursue and put resources behind their discovery.

It's just unknown territory and especially in the food industry such unknown or unchartered territory makes every member of management nervous. I have written this before: the food industry is historically risk averse and, because of the comparatively low stakes of the outcome, this is even understandable, at least in a historical context. This becomes different when we look forward and what the future might have in stock for the food industry at large, and this just in the limited first context of growing world population, climate change (man made or not bears no importance), limited agricultural surfaces and, last but not least, scarcity of water in areas where the majority of present-day's agricultural output happens.

6.2.1 No risk, no fun, or else?

And then there is this other saying that dates back but is equally used and loved now: "no risk, no fun." I am not suggesting that the driving force and motivation for any business should be fun; on the other hand if the consumers don't sense a pleasure and fun part in their food and beverage products, something likeable and emotionally engaging, which even might bring a smile on their faces, then no product might deliver and bring success to the company. Although "fun" is not a driver; it should be built in the product and especially service for every consumer. You might argue that I exaggerate, and you may even be right, to some degree at least, however I can say with some authority that without any element of fun and some element of playfulness nothing really works properly and successfully. And that is especially true for our food and our beverages.

6.2.2 What's architecture got to do with this?

I started this section by suggesting that the consumer is king (or queen), however, I would like to challenge this with an interesting example from the world of architecture. Let me use some of the principles of the world famous architect Frank Lloyd Wright as support for my challenge. Wright was probably a genius in his own right and his architectural feats are legend. Let me use two of his many architectural principles to state my point, namely that the consumer (the user in this context) is not always on top of the equation. Firstly, he had this important principle of keeping any entrance hall to a house or a condominium as small as possible, thereby forcing the owner or tenants, and especially visitors, to enter into the main part of the house, the living room as expediently as possible and not spend too much time at the entrance. This was and still is a clever way of "manipulating" the consumer. The consumer is definitely not king but is forced to behave in specific, predesigned ways.

The second of his many principles, which I want to use here, is seemingly even more radical. He typically dimensioned and placed windows with a view in ways that such a view could only be achieved by sitting down first. The upper limits of windows were low and the view could only be enjoyed in a seated position. The idea behind this was simple: he wanted people to rather sit down and enjoy the architecture, the room, the view, and everything that comes with it in a relaxed and seated fashion. Again, the consumer is forced to accept a situation that is imposed and not based on his or her free will and decision.

6.2.3 In search of the ultimate answer

I am convinced that such examples from a totally different perspective than the world of food can teach us important lessons to be applied in our own environment: how can we constructively "manipulate" consumers to make choices that result in winning situations for both parties? This is now the underlying question of this book, how can the R&D group contribute in decisive and successful ways? The answer to this question is really the "ultimate answer" in almost Douglas Adams' fashion. It's a really *big* answer and it's not an easy answer, leaving much doubt, uncertainty, and distrust vis-à-vis those who believe that they may be able to give the answers, especially the R&D group in every food company.

The question should be put forward in a different fashion: what is the Wright-type of approach that the R&D group can define and apply to put the consumer on a pedestal but not necessarily in the driver seat. And, as already previously mentioned, it is not only about the consumer but increasingly also about the customers of the food industry, the trade, the distributors, the super markets, and online distributors such as the Amazons of this world. E-commerce will probably substantially gain in importance and size and more and more consumers will turn to online shopping one way or another also for food and especially for beverages. Only a few years ago, grocery, although relatively small in absolute numbers compared to other businesses, was one of Amazon's fastest growing division, and the trend is still going strong, given that new players such as Walmart entered this segment.

6.2.4 Emancipate from the consumers!

This has important consequences for new product, process, and packaging development for the food industry at large and enormous challenges not only for the business side of the company but especially the R&D group. Such challenges include new types of packages with improved physical and barrier characteristics, new processes that take modified and potentially new distribution channels and timelines into account, and potentially search for and introduce new types of ingredients that can take "the hit" of a potentially less controlled distribution supply chain. Overall, it finds and defines products and ranges of products that are still perceived healthy, nutritious, safe, and tasty despite the fact that they arrive by mail carrier or, why not, drones that drop off your order on your front door or porch.

Again, it is the justification of "good for the consumers" that will bring about change, first to the food company, any food company, at large and in turn to the R&D group. It's this absolute commitment to the consumer that will drive change at any level. And change has already begun, slowly though yet steadily. Large food companies have taken longer for such change than smaller ones, and large distributors of food have taken longer than the smaller, more agile ones. It has to be said, though that there is of course more at stake, good or bad, in absolute numbers for the larger companies and distributors; therefore, it is only understandable that their level of hesitation is higher when it comes to readiness and preparedness for change. From personal experience I know for a fact that the majority of members of the R&D group in any food company is ready for change, although they may not always be well prepared. Why is that so? Well, simply because the composition of competencies that are believed to be necessary to solve the issues of food and beverages for the future is largely built on the recognition of issues from the past, at best the present and typically help to reply to these.

To take up this chapter's topic of a consumer perspective for changes in the food industry and especially in its R&D groups, I propose to shift the emphasis from:

- Outspoken and recognized consumer needs to unmet needs.
- Traditional, ingredients and processes based new food product and process development to food- and nutrition knowledge-based services to customers (the trade in the middle) and consumers (the end user).
- The usual renovation to real innovation, as difficult that may be not only to really discover but to make this understood to the management of food companies, which is traditionally—and I have said this a few times already—rather risk averse when it comes to going into new and unchartered territories. Renovation, after all, is a safe bet and not much can go wrong, and if something goes wrong, it's less costly and less of a negative shade falling on the management that has asked for the renovation project to be carried out.

FROM	TO
Recognized, met need	Unmet needs
Product and process	Knowledge and service
Renovation	Innovation
Small baby steps	Giant leaps
Risk aversion	Risk taking

Figure 6.4 From old to new, from low to high efficiency, from low to high added value.

- Going forward in small and safe steps to daring the giant jumps that most of the members of the R&D group of the company are ready to take, because that's what they love to do: discover totally new things and risk to fail on the way.
- Note: I agree that not every scientist, engineer or technologist is ready to take the risks, but I know from personal experience that it's a majority; and that's a great start.

Figure 6.4 illustrates this pathway from old habits and approaches to new ones.

6.2.5 I think we may have the wrong people, oops!

The real issue here is that because of having only looked to the consumers of the past and at best the present, the typical composition of the know-how portfolio in any given food company R&D group is potentially not up to the task. Not because the people are not willing but because they have the wrong background, which, by the way is not their fault. Such highly specialized employees were hired by higher echelons in the company, and they seemed to know what they were doing. Or did they not? Well, it's of course not that easy and it's a general dilemma that every R&D group in probably every industry will find itself: do we have the right people for the tasks of the future? And the answer, most likely, will always be, maybe? There is no easy way out and there is no single answer or recipe to solve the riddle. There are maybe some tools that can help, especially when we bring the consumer back into the equation.

6.2.6 Observation and smart conclusion: Two successful siblings

The simplest tool of all is the recognition that everyone is a consumer of some sorts, especially when it comes to food and beverages: everyone eats and drinks, at least as far as I know. Especially employees in the R&D organization of a food company are often told that the only valid consumer insight there is comes from the business side of the company, especially the marketing group. They are the real savants and they have the authority to tell everybody else in the company what the consumers really want, they have the quasi-absolute truth when it comes to consumers. With all due respect, but this is nonsense. Everyone who consumes and who observes himself or herself or others around can gather extremely valuable consumer insight, which can become extremely useful for any future product, process, packaging, or service development. Smart observation is of course only one part of the equation, the second one being smart analysis and appropriate conclusions based on observations.

A great tool that I have personally experienced quite a few times is the type of observation that runs under the general definition of video ethnography, filming consumers, with their consent but with a kind of hidden camera so that they can behave in the hopefully most natural ways when it comes to opening any type of food or beverage packages. Watching the outcome of the filmed observations can often be entertaining, even funny. However, the goal is not to mock the test subjects and have unjustified fun, the goal is to come to the right conclusions for future development and, most preciously of all, yet rarer, to discover still hidden and extremely valuable unmet consumer needs to pursue in future development projects.

The following definition of video ethnography complements the short discussion:

Video ethnography is the video recording of the stream of activity of subjects in their natural setting, in order to experience, interpret, and represent culture and society. Ethnographic video, in contrast to ethnographic film, cannot be used independently of other ethnographic methods, but rather as part of the process of creation and representation of societal, cultural, and individual knowledge. It is commonly used in the fields of visual anthropology, visual sociology, and cultural studies. Uses of video in ethnography include the recording of certain processes and activities, visual note-taking, and ethnographic diary-keeping.

Video ethnography involves:

- *Observation, including extensive filming of practitioners,*
- *Allowing practitioners to view the video recorded material and reflexively discuss their practice,*
- *Transforming practice through practitioner led change, and*
- *Building the capacity for the ongoing and critical appraisal of practice.*

Video-ethnographic methods seek to foreground practitioner knowledge, expertise, and insight into the dynamics of their own work processes. This is achieved by first talking with practitioners about their work and organizational processes, and by seeking an articulation of the social, professional, environmental, and organizational contingencies that both enable and constrain their practice. By allowing practitioners to discuss their practices in response to video footage clinicians and researchers gain insight into areas of practice that may be benefit from redesign. Video ethnography is contingent on the researcher gaining the trust of practitioners, on becoming familiar with the site and on being trusted to be present at time and in places where critical conducts are undertaken.

6.2.7 Observation is king

I emphasize this type of consumer research for a simple reason: it efficiently combines techniques of ethnography (studying anthropology, sociology, and cultures at large), and which have a proven track record, with modern communication techniques and make use of video footage of consumer behavior through observation, analysis, deduction, and resulting action. It is a technique that is not used enough, especially not by the R&D groups of food companies, despite the great potential they represent.

Scientists and engineers rather chase after other, more technically driven consumer measurement methods such as eye tracking in front of supermarket shelves or measuring the duration how long each individual consumer spends in front of such or such product displays. And if there would be more to measure, they would go for it, for sure! It's simply got to do with the kind of curiosity, based on the kind of training that researchers have undergone; it drives them to measure, almost naturally. My point here is that simple and straightforward observation of what consumers do, how they behave when they are exposed to packages and products, ideally how they behave in their "natural habitat" at home, in their kitchen, at the dining table, can be so much more informing and powerful to simply discover and understand consumers. And yes, observing consumers/shoppers in a supermarket is part of this total observation approach, provided it is done with full knowledge of the observed and his or her agreement. The rules of the game are the same as for candid camera: the observed should be informed after the fact and asked for permission that the results of such observations may be used. It might well be that some of the observed may not agree but, from experience, the majority will.

6.2.8 What do I do with what I have seen?

The real challenge that comes afterwards is of course the question: what can and will I do with the results of such observations and learning? How can this be translated efficiently in new product, process, and service development? How can the researcher integrate these findings into his or her research project and translate this into successful and sustainable results? I do not have the answer, or rather answers to share with you, although I have written extensively about this topic of innovation in the food industry. The answers are touching mainly two areas: first the content of the innovation and second the successful driving of innovation projects to a profitable and sustainable end. I do emphasize *sustainable* here because new product or process development in the field of food is all about longevity and continued success. With the exception of the product group ice cream, which is more of a novelty area, almost all other food and beverage products are long lived, even if you see a "relaunch" in a new package, it might still be the same "old" product. In principle, this is not a bad thing, however, this rather easy way out by focusing on renovation rather than innovation, makes it most often too easy for the R&D group. They claim to be inundated by the mundane consumer-requested renovation projects and therefore have little or almost no time to rather focus on real innovation. That's a real shame because this is where the real value is that every R&D group, every researcher in the food industry can and should add.

6.2.9 Tell the consumers, don't let them tell you! At least try

This leads me to the apparently only conclusion that seems to be logical: dramatically change the approach that the food industry and especially its R&D group have to consumers: become an anticipator rather than a follower, be daring to propose really new things that make the consumers' lives better and which every consumer uses at least once a day, hopefully more often. The food industry researchers have to become daring and smart to discover and anticipate unmet consumer needs and become daring to pursue research and development in this area of innovation rather than to slavishly follow the ever similar results of consumer and market research. Such research most often is an alibi that one has done everything possible and all looked good, unfortunately the new product launch was still a failure. It's a cheap way out and is really the most often heard excuse to failed developments and launches: "we had great consumer insight!"

I do understand that making this leap of faith and following one's own instincts and conclusions based on personal observation combined with the skills and know-how that come with every person in the R&D group is still a risky business and is often doomed to fail. However, when you win, you win big and can make the proverbial "dent in the universe" for your company. I also do understand that such an approach cannot only be handled by project managers and experts and specialists in the R&D group, but this approach has to be deeply grounded in the belief and support of management that this is not only the more promising way forward but also the right one, the ultimately most profitable one. This involves of course a perceived or often real loss of control by management over the employees in the R&D departments and is therefore not liked by many managers. It's nothing for the faint-hearted managers in any company, and I have seen and experienced many of these, unfortunately. This has to change: management needs to support the approach toward real innovation of products, processes, or services, and that means that they have to change from a control approach to a trust approach.

6.2.10 The ultimate downturn: Administrative processes

Yes, there is this saying: "trust is good, control is better"; however, I believe that this is an excuse invented by the ultimate "control freaks" and should never be applied to the ways of how an R&D group in the company is managed. After all, the researchers were once hired because everyone involved in the hiring process believed in them and thought that they were the right people for the job; they were trained on the go and they are even paid a salary, typically monthly or biweekly, and are given additional monetary resources to do their job. So, they are worthy of all that and despite all that they still need to be pursued by all kinds of control mechanisms, red tape, rules and regulations and, worst of all, administrative processes. I do admit that I have a personal hang-up with such processes and those who know me and have worked with me have heard me poking fun of these processes and questioning their usefulness constantly. I am almost a bit obsessed with processes and have made it a point to discuss best ways how to ignore them, at best.

The worst part of this—what I would call "processitis"—is the fact that companies use them as an excuse to improve quality of work and ultimately quality of products for the benefit of the consumers. A former colleague of mine once calculated the time spent by researchers in their project work dedicated to such administrative process and came up, conservatively calculated, with a staggering 35 percent of time spent to fill in time sheets, project sheets, or other so-called quality related forms. Again, it is supposed to be for the benefit of the consumers but ultimately,

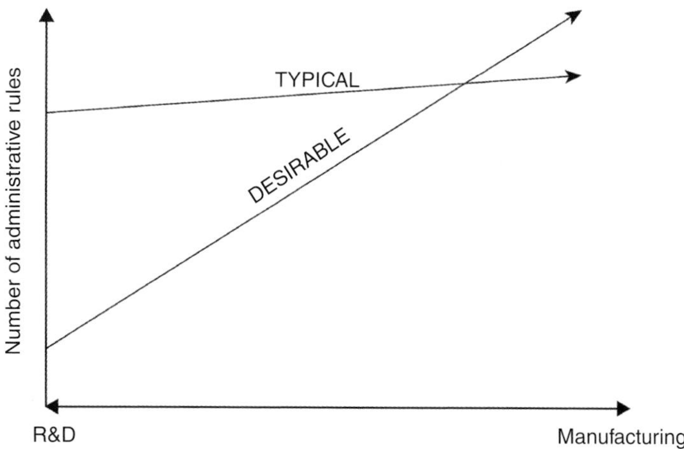

Figure 6.5 R&D and manufacturing in the maelstrom of administrative rules.

and that's the really dramatic part of this, it only makes the product more expensive for the consumer. This is a totally counterproductive situation and should stop immediately, if that was ever possible. And there is one more aspect to mention: the size of the organization has a lot to do with this processitis; the larger the organization the more you find of this, almost an economy of scale.

I do not deny that there need to be certain rules as to how one's work is done, following certain ethical, socioeconomic, and quality and safety standards and what might be perfectly appropriate, even necessary for a stringent manufacturing process is most likely not so appropriate for R&D projects. This is no excuse for becoming (or being) sloppy in one's work, it just is a call for becoming (or being) really sensible, especially when it comes to the environment of R&D and the consumers' impact as to how and why R&D pursues certain types of projects and new product developments and not others. This logically brings me to the next section in this chapter.

Figure 6.5 illustrates the desirable and the real relationship between the amount of administrative rules and, as an example, two functions of a food company: R&D and manufacturing, respectively.

6.3 THE CONSUMER-DRIVEN FOOD R&D

The popular belief is that every R&D project and product and process development should be driven and supported by consumer insight. I have said this before: things are not as easy and straightforward as this, it's definitely more complicated, and I really intend to challenge this consumer-driven R&D forcefully and shine some light on this from various different angles. Let me take a few moments to discuss the notion of consumer insight and how it is seen by different parts of the business at large and the R&D organization in particular.

What is meant by consumer insight? Let me quote a few definitions that I found.

Finding a brand insight definition must start with the consumer. Consumer Insights can be described as "enlightening knowledge about the consumer." It is seeing below the surface, understanding the hidden needs/wants/desires of consumers. It is the "Aha" moment, marketers and advertisers dream of. Insights are the convergence of real consumer understanding with brand relevance.

6.3.1 The "a-ha" moment

Got it? It's the "a-ha moment" and the convergence of real understanding with brand relevance, and this is probably all right, all good, and all necessary and useful; however, there is much room for interpretation and speculation as to what the consumer really wants, needs, desires, which leads to a recognition of what's the best way forward to be successful with consumers. I am far from applying a Cartesian thinking hat here, but this sounds a bit iffy and maybe hopeful. It misses this leap of faith that I can actually perform even without any consumer insight, and statistically speaking, I might be as successful as with a whole collection of consumer insight information.

You may have heard this before, but let me repeat this anyway: consumer research could be compared to building a bridge over a ravine that is narrow enough so that could actually also jump over it, however risky it may be. Building the bridge (i.e., having a ton of insight) might get you safely over the ravine, but there might be no consumers waiting for you on the other side because they have moved on during all the time you were building (collecting insight that is).

Let me add another definition on customer insight, which in our general context is absolutely comparable to consumer insight and in the following is used synonymously:

Customer Insight is the intersection between the interests of the consumer and features of the brand. Its main purpose is to understand why the consumer cares for the brand as well as their underlying mindsets, moods, motivation, desires, aspirations, and motivates that trigger their attitude and actions.

Another definition of consumer insight is the collection, deployment and interpretation of information that allows a business to acquire, develop and retain their customers.

A consumer insight can be more precisely defined as: "A non-obvious understanding about your customers, which if acted upon, has the potential to change their behavior for mutual benefit."... First, such insight is "non-obvious," so it does not normally come from just one source of information and often does not come from just analysis or just research; rather there is a need to converge evidence to glean insights. Second, true insights need to be "action-able"; hypotheses, which stay theoretical and cannot be tested in practice are not insights. Third, customer insights should be powerful enough that when they are acted upon customers can be persuaded to "change their behavior." Just benefitting from targeting based on analyzing past behavior and assuming people will be creatures of habit does not reveal any depth of understanding them, certainly not insight. Fourth, to be sustainable, the goal of such customer change must be for "mutual benefit."...a key law for marketing today is "earn and keep the trust of your customers," which is achieved by acting in their best interests as well as the long term value for the organization.

6.3.2 Take the risk and become independent

What I like about this definition is the part on "non-obvious understanding about your customers, which if acted upon, has the potential to change their behavior for mutual benefit." It's especially the part about mutual benefits, and the well-known and overly used win–win aspect is important and has to be emphasized. Moreover, the aspects of "non-obvious" and inducing "behavioral change" are equally important aspects and are all the right things to say. Again, it's a lot of words describing the uncomfortable situation that consumers may act in totally surprising and unexpected ways, and despite all the information that we could gather on many of the aspects, our new product, process, or service developments were unsuccessful. That's the big risk and even greater disappointment that we experience despite all the more or less valuable consumer insight.

The real question is, how do you find compatibility between the prevailing opinion of the marketing group, the business at large, which believes in the value of consumer insight and the probably necessary position of the R&D group, which should approach new product and process development, even basic knowledge building, in more daring, forward-looking and riskier ways? Again, this is not to say that consumer insight could not be valuable; however, from my own experience, it can be hindering to go after real innovation or, at best, delays the development process because consumer insight is still being gathered and the development project cannot be started yet. It's a tricky situation for the R&D group because on the one hand, it is funded, if not to say, sponsored by the business, and therefore should follow the business's desires and ideas as to what and how and when to develop new products processes and services. On the other hand, however, R&D should be "emancipated" enough to take its fate in its own hands and daringly go after their innovative ideas and push forward to real innovation in any type of development.

6.3.3 And better back it up with successful results!

It's a difficult balance, and I can say with much conviction, that the answer as to where this balance swings—solely consumer insight–driven versus pursuit of own innovation ideas based on smart observations—mostly depends on the personality of the CTO of the company (or a comparable function). It appears almost too trivial and again, I truly believe that personality is the main, maybe even the only, explanation for the direction that the R&D group takes or can take. It goes without saying that the CTO is not the only player in this balance game, it takes others below him or her who can and will pick up the ball and take on the challenge of new product, process, or service development in more independent fashion or whether they are too strongly impressed by an overpowering influence of the business side of the company. If CTO and R&D management and the individual researchers—scientists, engineers, and technologists—hold together, the battle, if there should be one, is half won. The other half of elements required to win the battle entirely can be summarized in two simple words: successful results.

It's as simple as this: if you as an individual or a group can show winners and successful results; it is difficult to argue with the approach that was chosen that led to such results.

6.3.4 I want to play with my own toys and make my own rules

I will discuss and analyze the various aspects of consumer driven–R&D efforts and the balance between doing development work solely or largely based on insight from without versus insight from within. I do realize that people with a technical background are often (or always?) accused of loving "their technologies" so much that all what they want to do is to push these toward the consumers. I think there is some value to it; however, if we accept this statement then the same statement holds also true for the marketing community. They like their consumer insight so much that all they want to do is to push the results of consumer insight back to the consumer, and in the case of the food industry, literally into their mouth. I have written about the need to find the right balance between this "consumer pull" or new product, process, and service development based on consumer insight, and "technology push" or new product, process, and service development based on the know-how, instincts, and innovative ideas of the R&D community (Traitler et al., 2014). The optimum solution, like almost always, lies somewhere in the middle and depends on the situation, the type of product or process or service, the type of the consumer group, and its maturity and cultural background. There are many factors to account for and I can say with certainty that the balance between consumer pull and technology push is never the same.

Let me say this, however; R&D, in collaboration with the other parts of the company, especially manufacturing, marketing and sales, has the obligation to define and, more importantly anticipate the "future consumers" and especially their needs and desires. This extremely important if not most important task cannot be abdicated to anyone else, especially not to marketing alone or worse, to consultants and trend gurus. R&D has the major responsibility in this and will always be rightly criticized in case they fail to live up to this challenge and help steer the company to a successful—even more successful—future of the company. And, like so often, the winning team will be composed of many players and coaches, whereas the unsuccessful outcome may only have one looser, hopefully not you.

6.4 CONSUMER GROUPS: THE PUBLIC OPINION

Yes, there is such a thing like the public opinion in food-related matters, especially when it comes to dealing with the food industry. Most of the time, it has to be said, public opinion with regard to the food industry is not a good one. Lately it has been lousy, and several decades ago, it was even worse. Why is that so? Well, in my opinion (no pun intended), the industry has done a lot to deserve it and has done little to anticipate and prevent a number of incidents that helped public opinion to turn rather negative, sometimes demeaning. I do not want to list all the incidents—most of them related to quality and wrong or fraudulent ingredients—but there were quite a few in the last couple of years. It also has to be said that such incidents are a great platform for politicians and other bureaucrats to come out, show their faces, and hold important press conferences. I do understand that the same bureaucrats are responsible to help protect the public from any harm, and "bad food" can be such harm. There is nothing to hide here and those who do, do a big disservice to the food industry as a whole.

Again, without going into any detailed analysis of past food-related incidents and the, in my eyes, very often inappropriate approaches by the food industry to respond to such incidents, history has shown time and again that in most cases such responses were coming late, maybe even only after an initial denial and were trying to minimize the case. Almost always, management had to finally admit that there was something not right and had to catch up with repairing the situation. In my personal recollection, there was only really one case in which the industry (and I use my former company as an example here) has responded in exemplary ways and that was after the Chernobyl incidents end of April 1986, when an entire response team composed of many expert resources was established in the shortest period of time and thousands of analyses of especially milk samples were carried out to separate good from bad or affected from not affected by irradiation. The most astonishing aspect of this is the simple fact that the industry is apparently capable of responding in fast and efficient ways but is not always willing to do it.

6.4.1 Early warning is the name of the game

And there are not only the food-related incidents that haunt food companies in frequent intervals, there are many more pieces to this puzzle that forms the public opinion. One of the major pieces is any type of consumer group, pressure group, interest group, NGO, or media reports as well as perceived faults and trends. The power of the message that may come from such sources can be extremely strong and potentially devastating, especially if companies respond in wrong ways to justified criticism. So, how to deal with this topic in the best ways? Most companies have adopted a strategy of "early warning" and most of the time this works pretty well, at least when it comes to real incidents that may be linked to wrong ingredients, wrong labeling, and "foreign bodies" of any kind in the products such as ink and curing agents transfer through packaging. Such early warning systems work on all kinds of levels from strictly factual to simply conversational to outright gossip (what might be "cooking").

There are many input mechanisms to gather relevant information that may or must be used to bring potential future incidents high up on the radar screen of the company's relevant management for such matters and may ideally be prevented from happening at all. This whole activity could simply be called "anticipation and smart prevention." However, it requires two important ingredients:

- It needs the necessary expert resources, which are in charge of such anticipation and smart prevention.
- There is not only such a dedicated group working in isolation but the entirety of employees of the company function as extended arms and eyes and ears of such an early warning system.

It goes without saying that without a free-flowing exchange of information and perfect communication within the company, such a system may be doomed to failure and its protagonists may just serve as scapegoats accused of mismanaging the situation. Without efficient communication the entire setup may just remain an alibi exercise.

In actual fact, it is not complicated to manage such an approach and instill the desire to contribute in every employee of the company. After all, every employee is an expert in or specialist of some type of more or less complex know-how with direct relevance to the cause at stake; the R&D group has a prominent and almost leading role to play.

6.4.2 Oops, we got it wrong

When looking in a bit more detail it can safely be said that the most frequent causes of food-related incidents are related to quality and safety, and this not only includes the wrong or wrongly dosed ingredients but also labeling errors or other packaging-related defects such as imperfect seals or wrong packaging materials specifications for a particular type of product. All of this is avoidable! I do admit that it's easy to say and write and more difficult to execute in the extremely complex world of the entire value chain—rather *value maze* as I have termed it elsewhere (Traitler et al., 2014) - of the food industry. Even in smaller companies this is not an easy feat, especially when some parts of this value maze are outsourced, and for cost and efficiency reasons, the company has to turn to comanufacturers or copackers. This makes the whole situation even trickier and can turn the public opinion against the company at fault quickly, even if the error was committed at the comanufacturing facility. "The buck stops with the one who is ultimately in charge, and that's the food company under who's name and brand the faulty product was manufactured."

It is difficult to turn the tide around again once a consumer group or an NGO has picked up a theme, especially when there are good reasons to do so, but even if the potential incident was blown out of proportion. There is a German proverb that goes like this: "Once your reputation is ruined, you can live happily ever after." Although this saying might work for marriage swindlers, it definitely does not work for companies. There is hardly any more difficult to repair than a company's reputation! On the other hand, all these groups, especially consumer groups, have an important role to play in checking the food industry and making sure that nothing bad ever happens. In this respect, they should rather be seen as allies of a food company and not as enemies, for what they are often perceived. It is understandable that it is sometimes difficult to fully embrace the opinion of someone or a group of critics that is not necessarily one's own view of things; however, it would be to the best of the industry if they would exactly do this: embrace the opinion and findings of critics and try to solve any apparent differences and disputes in a good dialogue rather than by rejection.

6.4.3 Working together for the common goal: Consumer benefits

It is, however necessary that consumer groups and NGOs have an open mind too and see beyond their immediate successes in more successful fund-raising results for their cause. In the long run both parties, the industry as well as consumer groups and NGOs who take an interest in the food industry, have a common goal, namely to improve any bad situation with the benefit for the consumer in mind. After all, the consumer groups as well as the food company have the

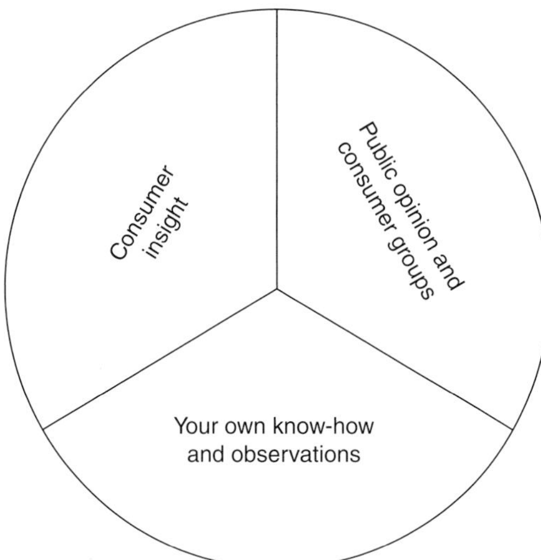

Figure 6.6 The main drivers behind R&D activities.

same interest of making any type of industrial food and beverage safer and more agreeable for the consumer. This also accounts for policies that govern and rule the food industry, which are to an important part influenced by the public opinion coformulated and shaped by consumer groups and food-related NGOs. Such policies should be meaningful and relatively easy and straightforward to apply so that there is no ambiguity and excuses for the industry to execute and the respective government organizations to control their implementation.

This in turn means that consumer groups and NGOs, the public opinion at large, can and must have a substantial influence on the kind of research and development that is carried out in a food company. This element together with the other elements such as consumer insight and own know-how and instinct are forming the backbone of why and what and how is researched and developed in the food industry. This is by far the best way to move forward in the food industry, and it should also be understood that each of these three elements may be of different importance at any given moment and for any given new product, process and service development.

Figure 6.6 illustrates the three main drivers behind the activities of R&D. It goes without saying that the main driver and reason for R&D to exist is the various needs defined by the business.

6.5 SUMMARY AND MAJOR LEARNING

- This chapter discussed and analyzed the critical consumer perspective for change of the food industry's R&D groups. It discussed various aspects such as what fast-moving consumer goods industry (FMCGI) represented and of what it is composed as well as the role of the consumer and how he or she can influence, even drive to some degree, the direction and ways that the R&D in the food industry operates. It also discussed and analyzed the role of public opinion, which groups are behind such public opinion and how it may be shaped and be used in profitable ways by the food industry by not falling for antagonistic positions but rather acceptance, dialogue, and collaboration with consumer groups as well as other shapers of public opinion with respect to food in general and with special emphasis on industrial food.

- The FMCGI is defined by manufacturing, marketing, and selling products quickly and at relatively low costs. Examples include nondurable goods such as soft drinks, toiletries, over-the-counter drugs, toys, processed foods, and many other consumables.
- The FMCGI's emphasis lies on speed, frequency, and cost, which are important elements explaining the type of pressure that is exerted and direction that is given to the results that are expected from the R&D group in a food company.
- Another key element in the FMCGI is packaging or the simple fact that all goods are packaged. This puts packaging almost at center stage, a place where it is typically not always seen, neither by the food company at large nor by the R&D group.
- Other key elements that are more consumer-driven were discussed such as increasingly low prices for food and beverages, and they should be fresh, nutritious, and safe. In addition there are trends for all natural, organically, and ideally locally grown and the shortest possible ingredients list.
- Food companies are typically risk averse and instead of daring risky product, process, or service developments they rather wait them out; someone else might want to go first and show the way and demonstrate that it is possible to achieve.
- Lean and efficient: the most important buzzwords in the food industry today. However, it is crucial for the industry to combine speed and efficiency of resource usage as well as execution with ample depth of the R&D process. Mutual understanding of the various partners in the company, especially also the R&D group are key to success.
- The best ways forward for the R&D group to reply to the "lean and efficient" pressure are: fewer and more focused projects—saying no to unreasonable and unrealistic requests from the business—be better prepared to anticipate requests from the business and optimize usage of available resources—increase usage of external expert resources—accept increasing constraints and turn them into a source of inspiration and increased creativity.
- The role of the consumer in the center was extensively discussed and it was emphasized that there is no such one person as "the consumer." The value and importance of "met" versus "unmet" consumer needs were analyzed and discussed. It was suggested that following only the met needs is the basis of most of the development work done in any food company and which is widely called "renovation work." Real innovation would, however come from following the more risky path of identifying and anticipating consumers' unmet needs. This would require some sort of emancipation from the consumer and moving toward a grown-up, risk-embracing group of employees and researchers in the food industry that is willing to put their own ideas and competences into discovering the yet unknown.
- The main consumer perspectives for changes in the food industry and especially in its R&D groups were discussed and analyzed. It was suggested that the emphasis has to shift from recognized to unmet consumer needs—from the traditional product based industry

approach toward nutrition know-how service to customers and consumer approach, from the usual suspect of "renovation" to the more unusual one of real innovation, and —from avoiding risk to embracing risk and being hailed for it, or at least complimented!

- Ordered to be more successful in recognizing and anticipating unmet consumer needs, the importance of smart observation was discussed and complementary tools such as for instance video ethnography were analyzed and discussed. The real challenge, however comes after the fact: what can and needs to be done with the outcome of any type of observation? How can the results be interpreted and ultimately applied?
- The biggest required changes for the industry, and especially the R&D group, that are largely based on consumers, their behavior and expectations are: anticipate, do not simply follow—be daring in going after really new things, even and especially if this would mean fewer and more focused projects—become smart enough to discover the unmet consumer needs, and break free from slavishly following consumer research put in front of you (if at all) by marketing.
- The role and counterproductivity of administrative processes in a food company were analyzed and discussed, and it was critically questioned whether the industry should make the consumer pay for the increased costs that such processes bring to any product or service. It was mentioned that in an analysis carried out in my former company it was estimated that approximately one-third of the time of a researcher in the R&D group is spent on nonproductive work such as filling in project sheets or other administrative tasks.
- It was suggested that degree of freedom that the R&D group in the company can achieve from being caught up in administrative tasks or other unproductive work largely depends on the role that the CTO (or comparable function) plays in the company. Strong CTOs who believe in the value of innovation will definitely serve the R&D community better in this endeavor.
- The role of public opinion largely shaped by consumer groups, NGOs, and the media was analyzed and discussed, and it was suggested that setting up anticipatory tools such as early warning networks can help to alleviate any type of crisis. Key to a successful and efficient outcome is collaboration and communication between the various parts of the company and especially within the R&D community.

REFERENCES

"Consumer insight," retrieved from http://360brandinsight.com.

"Consumer insight," retrieved from https://en.wikipedia.org/wiki/Customer_insight]

"Fast moving consumer goods," retrieved from https://en.wikipedia.org/wiki/Fast-moving_consumer_goods.

Traitler, H, Coleman, B, Hofmann, K 2014, *Food industry design, technology and innovation*, Wiley-Blackwell, Hoboken, NJ.

"Video ethnography," retrieved from https://en.wikipedia.org/wiki/Video_ethnography.

7 University perspectives for change to R&D in the food industry

Science is a wonderful thing if one does not have to earn one's living at it.

Albert Einstein

7.1 HOW DID WE GET TO THIS?

Let me begin by first introducing myself and then setting out my view of the current state of food related research.

I used to be a Chemical Engineering academic, but joined Nestlé on sabbatical in 2001. This involved moving from Birmingham, in the heartland of the (former) United Kingdom manufacturing industry, to Montreux, a beautiful lakeside Swiss town. As a chemical engineer working in a food and nutrition research center, I had the almost unique distinction of knowing next to nothing about food. This turned out to be something of an advantage and disadvantage at the same time!

I vividly remember early discussions with my new colleagues about various physical phenomena, which I generally believed that I understood from a generic science perspective. Seemingly this wasn't the case for food because my physics-inspired suggestions as to what I believed to be occurring were often met with the response, "aaah, but this is food," followed by a knowing shake of the head. After I while, I began to understand what was going on. Food science and food engineering have developed as relatively different and distinct disciplines from the so-called pure sciences and mainstream engineering. This is not in any way a judgment of the relative value of each because both approaches have strengths and weaknesses, but it is important to understand when discussing food-related research. I think a good analogy can be found with vertically and horizontally aligned businesses.

Food Industry R&D: A New Approach, First Edition. Helmut Traitler, Birgit Coleman and Adam Burbidge.

7.1.1 Why have "food science" and "food engineering" developed in parallel to mainstream science disciplines?

Food science and engineering are organized in a predominately vertically aligned manner. I have often wondered why this occurred, but I think part of the answer can be found by analyzing the strengths and weaknesses of vertically and horizontally aligned approaches.

The key advantage of a vertically aligned view is that the detail- and system-specific behaviors you research are almost certainly correctly dealt with because the system you are dealing with is the real system! Conversely, the disadvantage is that the system is a specific thing, so, in the lack of any simplification, it is difficult to find common features with other systems and learn from them without reinventing the wheel.

Figure 7.1 illustrates horizontal and vertical research based approaches to problem solving.

Classical science has taken the diametrically opposite approach and has worked with the simplest model systems possible to attempt to access the underlying general rules. Physics has recently taken this to extremes with the search for the underlying equation of everything and biology had the human genome sequencing effort. Doubtless chemistry could hold up similar examples, but I'm not a chemist so I won't try and guess what they might be. Clearly the advantages and disadvantages here are completely the opposite to those of the vertically aligned view in that a generally applicable answer is obtained, but that answer will be subtly (or overtly) incorrect in any real situation other than almost laughably simple ones. This is not to say that this view is worthless—I believe exactly the opposite—but it's clearly difficult to apply this kind of learning to solve "dirty" real problems.

As we have seen, food systems are some of the most complex systems we could imagine. Generally they are horribly out of equilibrium, making it difficult to apply classical equilibrium thermodynamics, and biological in origin, making for seasonal variations and all kinds of other uncharacterized hidden details, which may or may not play a dominant role in any particular observed behavior. When we consider that almost all classical chemical engineering is based on

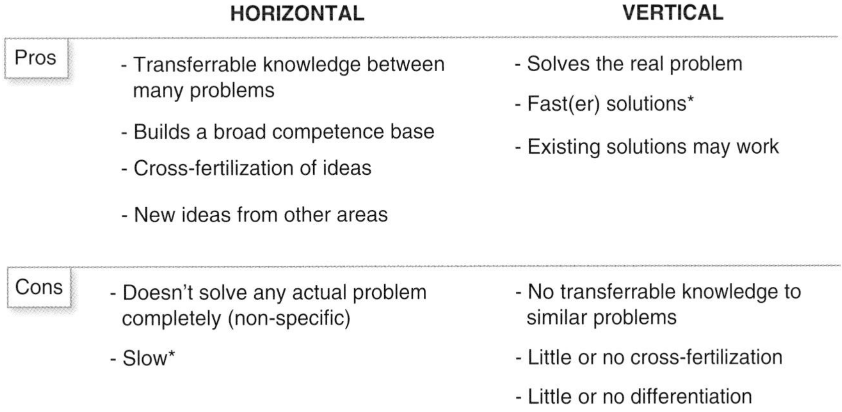

Figure 7.1 Pros and cons of horizontal and vertical approaches to solving problems via research. Note that the asterisked points are often strongly weighted when choosing which approach to follow.

chemically simple, near equilibrium behavior, it becomes relatively clear why food engineering has developed to a large extent as a parallel discipline. The previous discussion also goes a long way in explaining the parallel development of food science as a separate discipline.

7.1.2 Why does industry sponsor research

Later on we will address the question of research motivation from the perspective of a university, but first it seems appropriate to ask why does the food industry sponsor research? Actually, many industrial organizations will have research capability of their own, so, in this context at least, *research sponsor* is taken to mean "industrial sponsor of outsourced research." It seems fairly obvious that commercial concerns are essentially money-making enterprises, and as such, must generate income for either shareholders or owners. This suggests that the answer to the question is industry conducts research to be able to make money.

In the broadest sense, this is a true statement, but, if taken too literally, it misses out a lot of motivations for external research collaborations. Figure 7.2 outlines some of the wider reasons that industrial concerns might wish to carry out external research.

Let's consider these in a bit more detail.

The lack of a particular specialist is the most obvious and probably the most common reason for collaboration. Clearly it is not economically viable for a food company to maintain a full-time specialist for any science problem that could arise, particularly if questions in a certain area are fairly irregular. Deep specialists are, by their nature, usually quite narrow in their focus and competence, so it is difficult to find work for these kinds of people in between answering really important problems that arise on an irregular and unpredictable basis. In fact, this is one

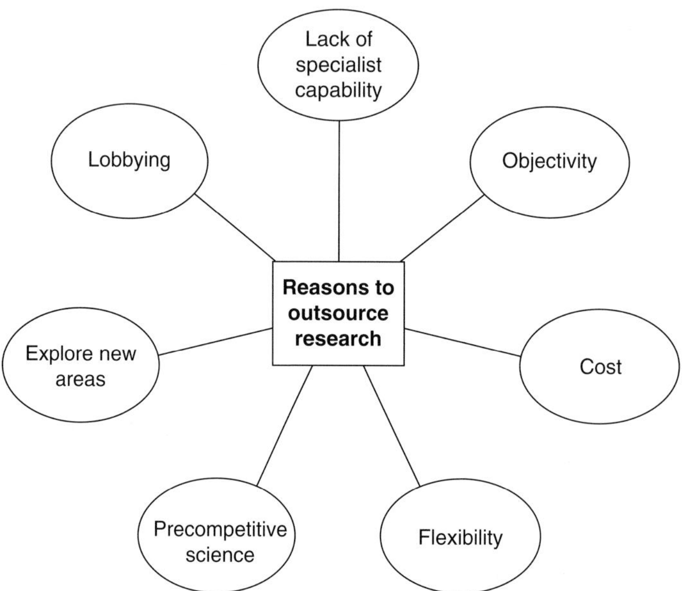

Figure 7.2 Some possible reasons to work with academic partners.

of the key issues with maintaining industrial research organizations, and the choice as to whether or not to maintain expertise internally will usually come down to a judgment call on how differentiating the application of that particular knowledge is likely to be. As mentioned, the flipside of this is that you may need to have some highly qualified people "sitting around" waiting for the right moment to make an enormous impact. Given the relative salaries of R&D specialists and business people, maybe this isn't actually that significant an expense after all? However, in these times of technocratic management by Excel tables and trying to drive efficiency via hours in projects, this can create some tension.

Moving on to the less obvious elements, leads us toward thinking about the objectivity of research. This can be either real or perceived. I guess you could debate whether there is a difference between perceived and actual bias, but in most practical cases, it doesn't matter. There are many situations where the credibility and objectivity of research is likely to be, or risks being seen to be, influenced by the interests of the company. To give a food-based example, imagine that an internal research project found evidence that eating as much refined sugar as you like does not cause obesity. This is a finding that would fly in the face of prevailing public and regulatory opinion; hence, however good the science might be, it is not likely to be taken seriously or at best be ignored. From this perspective, sometimes it is better to make sure that there is at least an external, independent component to this kind of potential conflict-of-interest research. One can of course make the observation that an academic lab sponsored by industry is hardly independent. However, it is at least one step removed from full industrial control, and as such, is instantly more credible. Sometimes, for particularly sensitive areas, this is not a sufficient level of distance and apparent independence. In those kinds of case, it might be better to involve a government agency directly in the funding of the research. This starts to overlap with, and invoke concepts of, lobbying and precompetitive science, which also appear in the figure.

7.1.3 IP "there's gold in them there hills": The intellectual gold rush

Way back in history (around the 1960s and early 1970s), universities were somewhat idealistic, predominantly left-wing organizations (in the loosest possible sense) that, at least superficially, believed in creation of knowledge for the good of mankind. Looking back, this seems like a hopelessly naïve and amateurish view, in serious risk of exploitation by the "evil forces of corporate capitalism," and in fact, at least according to the universities this is what happened. Thus began the war over IP.

Although it is doubtless true that inventions that should have been owned by universities were, on occasion, ripped off, this was a fairly rare event. To be fair, it wasn't rare because of a lack of will to execute, more that the inventions themselves were rarely worth very much! This then is the point of dispute in the IP war, which continues to rage to this day.

Somewhere toward the end of the 1970s and early 1980s, universities, and possibly more significantly, governments, who owned and paid for universities, got the idea that IP generated significant profit and that they should get a "piece of the pie." You have to admit that this seems like an entirely reasonable point of view. After all, universities are expensive to maintain and clawing back some of that investment seems justified. Consequently, universities started to employ more and more lawyers, and created "technology transfer offices" charged with managing and exploiting IP. Again, this seems like an entirely reasonable approach, and a much needed professionalization of the previous situation in which individual professors hawked their own things. Unfortunately, as with most things that involve lawyers, this made collaboration slower, more complicated, more expensive, and sometimes, impossible.

The crux of the problem is then the basic economic judgment of the value of IP and providing fair recompense to universities for their intellectual input. Written like this it doesn't seem like it should be too big a problem until you consider that, at the point at which the contract is written, the IP doesn't yet exist! How then should we value something, which does not yet exist, and indeed, may never exist, at some (unknowable) point in the future? Recent experience with the banking industry suggests that we, as a society, are not good at this, and that human nature tends to overvalue things. The university-industry IP standoff is similar in that universities tend to overvalue 99 percent of their IP. The justification for this is that the 1 percent that they don't undervalue is the part that will generate significant income for them, so their position is understandable. Unfortunately, neither side knows which 1 percent (or even less) will have any value! The flipside of this observation is of course that 99 percent of the time and effort spent discussing (arguing) over speculative IP rights is a total waste of time. It would be nice to do a calculation of the total value of the total IP portfolio versus the cost of discussing it. Although I don't have any available data to make such a calculation, I think it is far from obvious that the 1 percent, or whatever it actually is, turns a net profit.

How then can we deal with this to build effective and mutually trust-based collaborations that benefit both parties?

One way is to work on so-called precompetitive research in which the output is in the public domain and usually shared by all partners. This is a great way of working, although it requires that any real problems and materials are abstracted to the point where the outcomes are no longer commercially sensitive. The knock on effect of this is that the industrial partner will need to carry out a significant amount of translation and internal development to use precompetitive project output in real applications. Evidently this is difficult, if not impossible for smaller companies that do not have internal competencies available to do this translation step.

Another, quite successful approach to this is the innovation partnership model, which is discussed in *Food Industry Design, Technology and Innovation* (Traitler, Coleman, & Hofmann, 2014).

An alternative, left-wing, view of this would be that actually paying for IP is just another form of tax. In this paradigm, the university would give away the IP to anyone based in the same country, in return for which the government would fund the university properly via the corporate tax that they gain on the profits of the company that uses the IP. The advantage here is that you don't need to value individual IP, or know which 1 percent (or less) is valuable. There just needs to be some level of agreement on the overall value of the universities to the economy. Whether or not this model could work in a global economy filled with multinationals is unclear. It would clearly face resistance in the United States (c.f. sales tax added at point of purchase argument).

7.2 THE "STATE OF THE ART"

7.2.1 What does the food industry know about academia?

In the next section we will look at what academics know about food, but first let's consider the food industry perception of academia. The food industry is comprised of those who work for it, and as such, the food industry view of academia is the view of academia of those people employed by the food industry.

In my experience, the vast majority of those working in food research have some kind of food or nutrition science background. The knock on effect of this is that the collective view of universities is seen through the lens of food science, food engineering, and nutritional science. As we have already discussed, this is a fairly narrow view of the potential available within the academic community, and, personally, I think it's an enormous missed opportunity.

7.2.2 Academics: Three different ones

Academics fall into a few broad classes of behavior when it comes to collaboration, personally, during my academic life, I have occupied at least two of these; my point being that context, personality, and interests of those concerned could influence which box I found myself operating in.

1. *Food scientists*: As discussed, food science takes a specific, and often molecular, view of the world. Food scientists are generally split into experts in protein, carbohydrate, or lipids and will often have an appreciation of the nutritional aspects of their chosen material. This has advantages and disadvantages as previously discussed, but to generalize, they are great people to ask if you have a specific problem where the subtle details are dominant. For example, suppose you wanted to understand the degradation of protein during a heat treatment? This is a thermal chemistry problem that is clearly dependent on the specific chemistry of the individual proteins, so a food scientist specializing in proteins would be an excellent choice of collaborator. Conversely, imagine that you wanted to find a replacement drying process? In this case, an engineer of some kind would likely be a more appropriate choice.

   These two examples are fairly extreme, but there are other situations that are much less clear-cut. One such example from my own experience relates to how to address the question of fat replacement. Should one ask the lipid specialists to find a replacement for a lipid? Lipid specialists will, on the positive side, have a deep understanding of the properties of the lipids that need to be replaced; however, they will have little knowledge of things that are not lipids that could act as potential replacements. In practice, the best approach, would probably, as is often the case, be to build a multidisciplinary team to tackle the problem.

   The major potential downside to collaborating with food scientists, as with all specialists, is the tendency for conventional answers to conventional problems. Given that food scientists are aware of most existing food processes and products, it is likely that they will, knowingly or otherwise, carry a certain amount of dogma regarding how the process/product works and behaves. This is certainly not a bad thing, but will bias thinking toward "known" solutions, which may or may not be a good thing depending on needs and acceptance of risk or otherwise.

2. *Non-food scientists*: This is a bit of semantic nightmare, so let me begin by defining non-food scientists as those scientists (distinct from engineers) who work in "pure" disciplines (e.g., physics, mathematics, biology, biochemistry, etc.).

   Most of these people will know little or nothing about food and food processing (at least from a professional perspective). This observation is really the crux of the difference between them and food scientists and has positive and negative aspects.

   First, here is the positive aspect. Given that these individuals have no real specialist knowledge or experience related to food, they don't have any dogma or preconceived ideas either. This frees them up to think about food and food processes in a completely fresh manner, which can lead to potentially paradigm shifting innovation. Let's be clear that they have just as much dogma and preconceived ideas as "food people," it's just that it's not related to food!

   Nevertheless, this sounds exciting and motivating, but there are still two significant negative aspects to consider.

   The first, and most obvious hurdle is that of translating or framing your problems into a language that they can understand. This translation is an extremely skilled job and takes a lot of time and two-way discussion from a scientist that understands both the food science and the pure science aspects of the problem. These kinds of individuals are not that common in the food industry, so this can be a significant limitation, particularly for small enterprises without large associated research organizations.

   The second, and less obvious hurdle is that these pure scientists may not be particularly interested in solving your problem! That might sound like a brutal statement, but consider how these academics are rewarded. Generally, academics build their reputation by publishing impressive and highly cited articles in high impact factor journals, and journals that focus on food tend to have low impact factors when compared to more mainstream pure science publications. This means that pure science academics are unlikely to want to publish in food science–focused journals, and consequently they may wish to use model materials rather than food.

   It is unfortunately true that there is a certain amount of snobbishness regarding the pecking order of high science, and in my experience it runs:

   pure science > applied science/engineering > food science/food engineering

The reality of this is well observed in the TV comedy show *The Big Bang theory* in which the pure theoretical physicist looks down the experimental (applied) physicist; both of them looking down on the engineer (despite the fact that he becomes an astronaut)!

I have observed similar prejudice in real institutions. For example, an ex-colleague of mine was extremely successful at collaborating with industrial partners, but the work was often disparagingly treated as somehow dirty by many of his peers.

Figure 7.3 Science versus engineering. Image courtesy of Brian Edwards - http://user47329.vs.easily.co.uk/ifference-science-engineering/

In summary, the effort and skill and time that is required to effectively engage is high, but returns and paradigm shifting solutions are most likely to come from this approach, although they will often require disruptive change to implement. Are you really ready for disruptive change? Is your management? If not, then following this approach is unlikely to make you new friends; but if they are then the sky is the limit! Spectacular failure is also, of course, a possible outcome.

3. *Engineers (food or otherwise)*: Engineers really want to solve your problem. They really, really want to solve your problem! In fact they want to solve your problems so much that they may try to solve problems that you didn't even know you had! (I'm not joking; I've done this myself!).

   Engineers like nothing more than building and optimizing things so that they work "better" and are usually quite pragmatic about how to achieve this. A well-known motto that gives an insight into the engineering mindset runs as follows: "if it doesn't work, try hitting it with a hammer; if it still doesn't work then find a bigger hammer!" We engineers are generally more interested in the end result than in how elegant the solution might be. Figure 7.3 depicts this traditional field of tension.

   Once again, this is both good and bad from the perspective of getting help.

   The good side of this, is, as already stated, engineers really want to help you and really want to get something implemented and working at an industrial scale. On the face of it, this seems perfect, so what's the problem?

   The difficulties are generally related to timeframes and information transfer.

   The first difficulty relates to compatibility of timelines. Universities are first and foremost educational institutions, and as such, have an obligation to train students. From a research perspective, this will generally mean postgraduate students and will usually require a coherent research effort on a single topic over a three- to four-year period. If you have a process that isn't working correctly, typically an engineering question, then the financial drive to solve the problems is usually urgent.

An alternative approach on the university side is to engage a postdoctoral researcher on a short-term contract. This is more expensive, but the researcher is likely to be able to make a contribution to finding a solution much quicker than a student, who would need to be trained first. Time frames here are more compatible, but good post-docs on short-term contracts are likely to start looking for their next job six months before the end of their contract (they are rarely permanent employees of the university) and can leave at the "drop of a hat," leaving your project unstaffed. It should also be noted that universities will rarely have (good) post-docs just "lying around" waiting to staff your project because most researchers are on so-called soft money and do not have permanent contracts. This means that recruitment time should also be budgeted for when post-docs are used.

The second difficulty that merits a discussion is that of information transfer.

To find a solution to a problem or improve a process, an engineer will require a lot of highly sensitive process data, and as a consequence, will develop an excellent understanding of how your process operates. Process subtleties and operating conditions are the bedrock of commercial advantage, and as a consequence are unlikely to be frankly and openly communicated to an external person. This is particularly problematic when the external professor is likely to work with competitors on similar problems.

A third issue, which occurs to me, is that of duplication of skill sets between academic engineers and research engineers in large companies. I have had a number of discussions over the years with academic engineering colleagues about how we could collaborate, and many of these have foundered on the problem that they have essentially the same skill-set and motivation as I do. This leads me personally to work more as a translator of problems to pure scientists because this is more likely to avoid skill duplication. As mentioned, this is only really works for companies with large research operations; I believe that engineers are probably the best choice of collaboration partners for smaller enterprises, and if need be, they can also take on the role of translator.

7.2.3 Nutrition, medical science, claims, and regulatory bodies

Thus far in this chapter, we have mostly considered the physico-chemical side of food in some detail, but another, equally, or perhaps even more, important aspect of food is its nutritional function. If you consider the motivation to eat, then the top of the list has to be nutritional function. You can make arguments that pleasure, social context, and so on, modulate the desire and the quantity consumed, and these are undoubtedly valid, but at the end of the day, in simple terms, you eat or you die!

So, having established that nutrition is essential to foods, we should at least discuss it in the context of university led research. Nutrition is, as with the physico-chemical side of foods, a deeply multidisciplinary subject, and as such is often attacked by researchers with differing backgrounds and foci. Consequently nutrition can be, and has been, viewed from the perspective of dietetics, metabolism, disease, epidemiology, sport, pediatrics, obstetrics, geriatrics, and many others. Given that there are so many actors involved, nutrition can mean different things to different people, and this can make the situation extremely complex from the perspective of finding consensus. Figure 7.4 illustrates suggested phases of nutrition research. As such, these

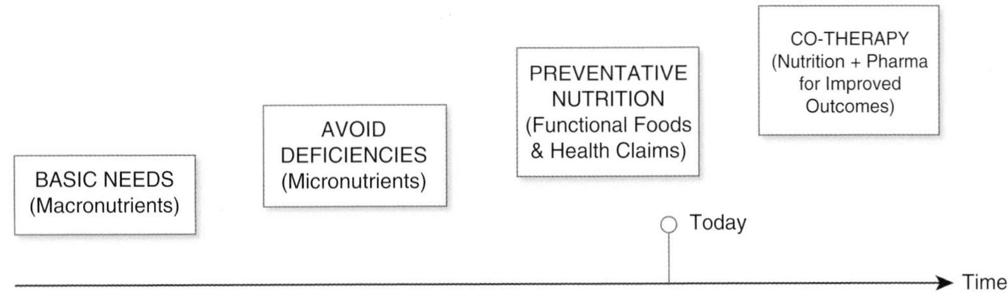

Figure 7.4 Conceptual phases of nutrition research.

phases are rather universal and not necessarily only to be found in academic research but equally in the larger research and development departments of food companies.

Phase 1: As basic energy (macronutrient)

This gravitates around the need to provide enough basic calories and protein to survive. This was the prevailing model before the early 1960s, by which point the food industry, in the developing world at least, had evolved to the point where providing enough calories to survive was relatively non-differentiating and basic processed food started to become increasingly commoditized. The developing world is at a point where this is now predominately true, although there are some areas where this is still the key nutritional driver. If we widen our definition of nutrition to include clean drinking water, then this scenario remains rather prevalent in quite a few areas.

Phase 2: Maintain metabolism (micronutrients and essential minerals)

Once basic calorie needs are satisfied, the next level of evolution is to try and address deficiencies of micronutrients, vitamins, and minerals that are essential elements of a healthy metabolism. This is an interesting one, because one could, and many do, argue that most diets before processed foods were to some extent "balanced" from this perspective and therefore healthy. I guess the validity of the argument depends on food availability and which population group precisely you are arguing about. But I think it's fair to say that there was a certain evolutionary pressure for the human organism to develop to be as robust as possible with respect to the prevailing diet. The problem comes in the definition of what that *diet* realistically comprised over evolutionary time scales, and how transferable any conclusions might be bearing in mind that the human life expectancy for the vast majority of history was sub 30!

Anyhow, it is beyond debate that many processed foods are missing certain of these micronutrients as an inevitable consequence of processing and standardization. This has led to a number of streams of solutions: supplements, fortified foods, and "natural" foods.

Supplements are essentially pills that deliver a high concentration of one or a blend of particular micronutrients to attempt to provide missing micronutrients. There is a lot of money to be made here, but the key problem relates to bioavailability of the active compounds. It is well established in the scientific literature that in many cases the fraction of these supplemented ingredients that are absorbed by the body is extremely small. The joke goes that "buying supplements is just an expensive way to supplement your urine!" (I'd like to say "piss your money away," but I guess we can't say that!)

Fortified foods are kind of an oxymoron in that something good is removed from a food and then added back in afterward. Although it may be technically challenging, surely it would be better not to remove it in the first place!

Which logical process leads us to "natural" foods, which ironically represent the antithesis of food processing! Hence the situation that we currently experience is that some food materials are actually more valuable before you process them, and that, by processing them, the value is actually reduced!

Phase 3: Preventative health (functional food—nutritional claims)

Having arrived at the end of phase 2, the processed food industry found itself with something of a problem. Processed food was heavily commoditized, and the strongest premiumization driver was to not process at all! This is a bad place to be from a commercial perspective, so a new paradigm was required to move forward. In the western world, this point was reached around about 1990–1995, which, coincidentally coincided with the peak of the pharmaceutical industry making a commercial killing. Logically enough, those running the food industry looked toward the pharmaceutical industry for a possible model to allow new, innovative, differentiated products that could drive growth and margins. Enter "functional food," and the concept of foods driving health benefits.

Superficially at least, this seems like a great idea. Foods that act like pharmaceuticals would be strongly protected by large organizations, command high premiums and would require significant, if not immense R&D investment to develop. The reality was unfortunately rather less wonderful. There were at least a couple of reasons for this. First, pharmaceuticals generally work by delivering a controlled "overdose" of and active material to drive a benefit. Food as a matrix is consumed in many different ways by different people. This depends on all kinds of uncontrollable factors, and as such, it is almost completely impossible to control dosing of the active material from a food product. Second, as already mentioned, consumers want "natural" foods, which is something of a tension with respect to consumer drivers for premium products. Third, the regulatory environment for pharmaceuticals is extraordinarily rigorous and complex, as justified by the margins. Even with high-end food margins, the investments required for clinical trials would prove to be too high to justify for all except a few SKUs of a few megabrands.

Phase 4: Co-therapy with pharmaceuticals for improved outcomes

Today we are entering the "postfunctional food" era, in which the classical view of functional foods as outlined is more or less dead in the water. However, that is not to say that nutrition and health are not strongly linked because I believe that they are. The question is how to better couple nutrition and health together to drive value. One emerging model for this is to think of food as complementary therapy that is intimately coupled with pharmaceutical intervention. For sick people that are undergoing clinical treatments, there is an opportunity for food and pharmaceutical companies to work together to execute clinical trials that demonstrate improved outcomes for coupled diet and pharmaceutical relative to pharmaceutical alone.

There are doubtless other models that can and will likely be explored in this sector, but for the moment, it is evolving fast and the final direction is unclear. One can ask the question whether functional food as a "preventative" measure for the broad population is likely to reemerge, but I think, for the reasons outlined in relation to phase 3, that the likelihood is no.

7.2.4 Getting money from governments via grants and awards

As previously mentioned, governments tend to pour large amounts of funding into universities. Why do they do this? The answer that politicians usually give when asked this question is along the lines of "to improve the competitiveness of 'reader please insert country here' PLC."

So how does that work? Well, there are two key aspects: training and innovation.

Training is obviously the teaching and development of competent, highly trained individuals that provide an attractive workforce for those private enterprises that power the economy (and pay taxes back to government). Universities play a key role in this through training of both undergraduate and postgraduate students.

Innovation, at least theoretically, improves margins and improves businesses; Universities are not businesses, but the idea is that linking the "innovation engine" provided by universities to the commercial engine of industry would create wealth for the country concerned. This seems like a credible argument, but large industrial concerns have (or more accurately, had, because we are talking about the latter part of the twentieth century when this approach started) their own internal research division, so probably don't really need universities. Enter the mythical small or medium enterprise (SME).

Speaking about SMEs was extremely trendy around the mid- to end 1990s and they were built up as the great, untapped wealth-generating potential of western society. In reality, they were the start-ups that never were. The idea was that you could take a small, agile business and tie all of that focus and drive to a great innovative idea or technology to build a great partnership. In the United Kingdom at least, this idea never really took off, and I think some of the reasons why have been highlighted previously. I think that the key point was the lack of a good translator on the industrial side, and an underestimation of the time and effort required to create a business from a good idea.

It's interesting to compare this late-twentieth-century view of a SME with the modern template for a technically driven start-up as outlined elsewhere in this book. I think the key phrase here is *technically driven* because this is different to the kinds of SMEs that I was dealing with as a consulting academic during the 1990s. *Technically driven* means that technically smart people are in charge and that they understand the differentiation and opportunity in a deep manner. These technically smart people are often ex-university professors and understand how to communicate with academics, which removes one of the largest barriers to effective collaboration (although the time scale problem remains acute). Clearly the modern IT or bioscience start-up is a different animal to the older businesses of the late-twentieth-century that we were discussing in the previous paragraph.

Anyhow, governments are keen to encourage interaction between industry and academia and to promote these kinds of activities they offer a number of different kinds of grants. These include

EPSRC/BBSRC CASE studentships (United Kingdom)
Swiss National Science Foundation (Switzerland): requirement to commercialize something at the end of the collaboration
Framework Program: European Union
National Science Foundation IIP Division, USA: various programs for Industrial Innovation Partnerships

Let's take the EPSRC/BBSRC CASE studentships as an example. There were couple of iterations in the details of how these worked, but in essence they were awards given to universities to sponsor PhD theses by research that required an industrial partner to pay a "top up" salary to the student. This was a good partnership on all sides: the university got a grant to engage a student and carry out research; the student got paid more than the basic grant; and the industrial partner got some say in directing the research topic toward something that they thought might ultimately be useful to them. Because of the improved salary, CASE awards tended to attract some of the best students, which was an additional benefit. Over time, industrial partners formed long-term relationships with academics, and this led in some cases to research programs. Eventually, the ownership of most CASE awards was moved from universities (who needed to find an industrial partner) to industrial institutions, who had a good track record in university collaborations. This was a far-sighted move by both the EPSRC and BBSRC and led to some excellent research collaborations over a number of years. Although this example is specific to the United Kingdom, I'm fairly sure that similar schemes exist around the world—some examples of which are highlighted in the table. The three-way partnership between government, academia, and industry is powerful indeed when it works well. Figure 7.5 depicts this well-established relationship.

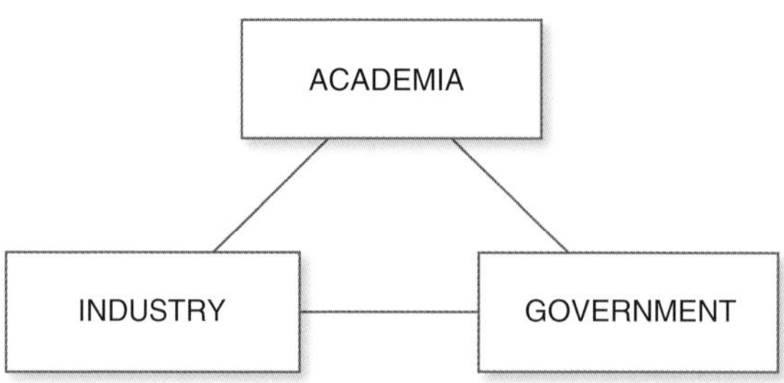

Figure 7.5 The "holy trinity" of research collaboration.

7.2.5 Academics as consultants

In addition to working via students and postdoctoral employees on a medium-term basis, many academics also offer private consultancy services to industrial partners. These are generally negotiated on a daily rate and the IP is usually ceded to the industrial partner with little or no negotiation. In fact, little new IP is expected to be created as a consequence of this kind of interaction because the contracted professor is usually offering an opinion on something and not doing new or original work as such. This kind of interaction allows for a valuable two-way exchange of information: contracting parties will often get new ideas and approaches to their problems and academics will obtain valuable insight into currently relevant industrial problems and trends, which can be used for fertilizing new ideas or projects and helps to keep teaching current and topical.

One thing that needs to be carefully managed in this kind of interaction is the potential conflict of interest that can occur when a professor doesn't immediately know the answer but can propose a further research-level collaboration that will address the problem in a deeper manner. In this case, the academic has a vested interest to propose a collaboration with himself or herself, which is often hard to resist, particularly if the problem is scientifically interesting. Many companies enforce a policy that forbids external contractors to be simultaneously employed as consultants and researchers to avoid this kind of situation, but it is far from universally the case.

7.3 WHERE ARE WE HEADING?

7.3.1 Reunification?

Having seen that there were justifiable historical reasons for having a separate food view of world, we should now consider whether or not this separation is still justified in the modern world. My personal view is a qualified no. That might seem like a strong statement, so I will now try and set out some of my arguments and justify this position, at least from the physical perspective!

Modern soft matter physics has advanced extremely rapidly in recent years, and tools and theoretical frameworks are starting to become available that open the doors to understanding food materials using the same approaches used for other complex, more glamorous systems. In fact much of these developments have been driven by a quest to "quantify" biology, and foods are, after all, usually based on (dead) biological systems.

Likewise, modern chemical engineering has moved away from bulk, continuous, near-equilibrium, product of simple chemicals toward high value, small volume, batch processes for specialties. Clearly these kinds of processes and systems are much closer to a typical food production, and as such, the knowledge ought to be more transferable.

7.3.2 Research as a marketing tool

Research is often used as a marketing tool in consumer industries. A particularly striking example of this is found in the washing detergent business, where commercials refer to mythical scientists, who are universally portrayed as benevolent geeks in lab coats. These scientists spend their days doing "science," which is represented as high tech and obviously beneficially to the consumer.

Another contrasting example can be found in the pharmaceuticals industry, which again is science-driven marketing, but from a different perspective. In this case, the target of the marketing campaign is physicians, and governments (to obtain reimbursement). The United States is a particular, and I think, atypical, case because there is a considerable amount of pharmaceutical marketing aimed directly at the patient or consumer.

The food industry finds itself somewhere sandwiched between the two extremes, but consumer drivers are different. When asked what they value for food products, consumers are driven by "naturalness," which is an interesting concept because it is essentially "anti-technology." Clearly there is some balance regarding the accepted need for safety; however, in the developed world at least, this is more or less assumed. So, do we have to conclude that science and food don't mix? I don't think so. I believe that the message should be interpreted as follows: consumers like to pretend that their food is natural, fresh, and unprocessed, but deep down they know that it isn't. They don't what to be reminded of this, nor do they want it to be made obvious, since highly processed food is, from societal norms perspective, synonymous with "junk food."

Foods with benefits are different; here science sells the product (e.g., infant formula, sports nutrition, ageing care, pet food, medical food). As discussed in nutrition, there are some areas where this science-driven marketing can work well and might even be seen as a prerequisite to play in the market. In general, these are markets where to manufacture has almost complete control of the consumer diet; this is usually necessary because nutrition works on a diet and not a product level, and the dosing of any active compound must be carefully controlled within narrow limits to simultaneously achieve functional benefit whist avoiding toxicity.

7.3.3 Crowd-sourcing solutions: Open innovation pros and cons

Modern technology is starting to change the research landscape and innovation landscape and one emerging aspect that is enabled by modern communications infrastructure is the private researcher. A *private researcher* is an individual who does research independently of, or has no affiliation to, a university or employer. Often these individuals are retired academics or industrial researchers who enjoy solving problems and have the time to think. There are a number of agencies that syndicate for want of a better word, problem statements on behalf of their clients to a so-called group of "solution finders" to get a crowd-sourced solution. In most models, the company that sets the problem statement will offer a prize for the best solutions, and this provides incentives to the pool of solution finders. Many large companies now use this approach and it is becoming increasingly popular.

The positive side of this is that it generates an extremely wide range of possible solutions, and that the solvers come from all kinds of different backgrounds an experiences, so the chances of something innovative emerging is likely high. The negative sides relate to possible misunderstandings based on poor problem definition, having to sift through a lot of potentially irrelevant (sometime borderline loony) ideas/solutions, and possibly the difficulty in building teams to deliver solutions in the manner that potentially could be achieved in a dedicated research organization. The competitive element could therefore be viewed as either good or bad. For sure, all of these problems are manageable, so I think that crowd-sourcing is likely to form one of a number of pillars of future corporate research organization.

7.3.4 Scientific publication in the future

Publication and dissemination of research via peer-reviewed scientific journals has always been a primary quality indicator of scientific excellence. It is also the primary mechanism of knowledge archiving, but things are changing rapidly. The disadvantages of printed journals are similar to those of printed books, and as such there is increasing pressure to find a better way. The high cost of purchasing an increasingly large number of diverse journals is coupled with decreasing library budgets and on-demand pay-per-view Web-based systems is also having a significant impact. In addition, government funding-agencies led by the National Institutes of Health (NIH) in the United States are increasingly demanding that publically funded research should be publically available. Given that the vast majority of published research is publically funded, journal publishers are coming under strong pressure to provide free access for everyone. Of course, this means that the traditional funding model for journal publishing is starting to collapse and that publishers need to find alternative income streams. In most cases, this means that journals increasingly try to load the publication costs on the article authors.

Despite considerable inertia, alternative modes of science dissemination are starting to emerge and gain traction. Most of these are Web-based with the significant benefit over printed matter that interactive dialogue becomes possible. Another benefit is speed of publication. One major concern that is commonly raised relates to archiving, but I'm not sure how credible this really is. In fact, the physics community has addressed these limitations for a long time through the preprint servers, where article pre-prints were posted for discussion before usually, but not always, printing them in classical journals. The largest of these is called arXiv, which is a strange name until you realize that X represents the Greek letter chi and everything sort of makes sense in a science humor kind of a way! A key limitation, or advantage, depending on your point of view, is the lack of any kind of peer-review process. This could have the consequence that quality control is lacking, although there is little evidence that this is the case, as with Wikipedia, where the technical articles are of a surprisingly high quality. Those of us that have experienced journal peer review firsthand know that it is far from a perfect process, so, I think, on balance the preprint server is a pretty good system.

The next step will surely be to remove the printed journals altogether, the only real question being how long this will take to play out and how will it be replaced.

Another recent development is ResearchGate. This is a kind of scientific Facebook/LinkedIn service, combining a Q-n-A feed, online CV, and self-archiving of publications. I believe that this kind of service will totally revolutionize the way in which academic knowledge transfer takes place. I think this is important because it offers a new way to share information and to build networks. This network-building function was previously fulfilled via attending conferences and giving talks, and consequently, is likely to have a knock on effect regarding the function that these events serve. Academic careers are becoming increasingly like running small businesses and entrepreneurial and brand-building skills are now more or less essential for success. Digital brand building looks to be the future.

7.3.5 A multidisciplinary future

One thing that seems abundantly clear to me is that the majority of innovation and progress in the future will likely occur at the interface between disciplines. Many areas of science are already extremely highly developed, but often narrow in their scope. As such there is enormous opportunity to put the pieces together from apparently unconnected chunks of research to shift paradigms. Universities were historically organized in an extremely vertical, mono-disciplinary manner, which makes cross-disciplinary working difficult. Actually, it's notable that some of the older institutions such as Oxford and Cambridge run a college system that encourages interaction between professors from different disciplines in common rooms and over meals. This system has its origins in the monasterial origin of learned institutions, but somehow, this template appears to largely have been ignored, when most of the first-wave of classical universities were created at the tail end of the nineteenth century.

Will this changing environment cause universities to reorganize themselves? I think it probably will, but I can honestly say that I have no idea how it will pan out.

7.3.6 How to collaborate better?

So, having outlined the past, current, and speculated about the future, it's natural to pose the question how can we collaborate better? I would propose the following three simple guidelines:

- We need to have specialists on both sides; we need to really take the time to understand our problem/opportunity
- Timelines: universities work in units of PhD or years of post-doc; often this is not compatible with short-term problem solving (urgent), so we need to adapt problems to timelines.
- Be clear up front on a reasonable definition of a successful outcome.

7.4 SUMMARY AND MAJOR LEARNING

In this chapter we have taken a look at the state of academic research in the food industry and how it connects or otherwise to industrial and government partners. The role of multidisciplinary teams cannot be underestimated in the future of driving innovative and hopefully profitable academic–industrial partnerships. The academic model is probably not well aligned with current industrial research needs; both sides will need to adapt their approaches in the future.

- Academic research in food science has developed somewhat independently and in parallel to mainstream classical scientific disciplines.
- This parallel development has probably occurred because of the complexity of food materials and the consequent problems of applying simple, equilibrium thermodynamic concepts that underlie much of classical physics and chemistry.
- Solutions found for simplified "model" materials are often difficult to apply to dirty, complicated, "real" food materials.
- Industry sponsors academic and other external research for a range of reasons. Typical examples might be lack of specialist internal capability, cost, flexibility, and precompetitive science to explore new areas.
- IP rights present a considerable hurdle to efficient and mutually trusting collaborations between industrial and academic partners.
- Valuing science outcomes and inventions from research that has not yet been carried out is extremely difficult to do in a meaningful manner. Nevertheless, universities tend to overvalue 99 percent of their IP when it comes to drafting research contracts.
- We need to find alternative methods of collaborating in an administratively and legally lighter fashion if we want to foster truly productive academic–industrial collaborations.
- Different kinds of academics will provide different kinds of solutions: Food science people will provide good solutions, but they will likely be quite conventional; non-food science people will provide things that may or may not be solutions but will likely be unconventional. Engineers will try to do your job for you.
- Nutrition is at least as important as texture, structural, and materials science from the perspective of food.
- Nutrition started out providing basic nutrients but moved increasingly toward trying to co-opt food matrices as a vehicle for health benefits. Unfortunately, outside of a few specific cases, mainstream consumers actively don't appreciate health benefits and functional food. There is a developing segment that views food as a co-therapy together with conventional pharmaceuticals.
- Governments see academic–industrial collaborations as a potential engine for economic growth.
- Governments provide attractive cofounding support for academic-industrial collaborations, which, when well used is an excellent resource that satisfies the needs of the three partners.
- Academics usually provide independent consulting services, which can sometimes be a quicker route to finding a solution or guidance related to a specific problem than a full blown university research collaboration.
- Physics, chemistry, and biology are beginning to reach a level of complexity that is sufficient to address some real, dirty food problems. Reunification of the disciplines seems like only a matter of time.
- Research makes an excellent marketing tool for that segment of the food industry that is science aware. However, it is likely exactly the opposite for the vast majority of food products.

- Science publication is changing fast. This provides opportunities and challenges in terms of how to disseminate results. From an industrial perspective, this may drive a need to be more open with the outside world.

REFERENCE

Traitler, H, Coleman, B, Hofmann, K 2014, *Food industry design, technology and innovation*, Wiley-Blackwell, Hoboken, NJ.

8 Industry perspectives for change to R&D in the food industry

While some people may think being a chef only entails making enticing dishes and pushing the culinary boundaries, being a part of the food industry involves much more. Without food, we cannot survive, and that is why issues that affect the food industry are so important.

Marcus Samuelsson

8.1 A TYPICAL FOOD INDUSTRY SET-UP

Most food companies manufacture a more or less large number of food and beverage products, which are based on their own brands. This is at least true for what could be called the "branded food companies." Another type of food companies works more in the background and acts as a co-manufacturer for what typically is called "private label" companies. The entire definition is a bit artificial because every brand for every product is typically privately owned. Historically, there were situations in which governments, typically dictatorial ones, owned the rights to brands and would therefore, by definition, own public brands and would represent a "public label" company. You can see where this leads us, namely rather nowhere; however, the distinction between branded and private label does exist and is widely used in the industry. The distinction goes much farther: a branded food company would never (or rather almost never) manufacture for a third party, even if their manufacturing lines would allow this, for both technical and availability reasons.

Branded companies would just not do this; they fear that it's bad for their own brands if they were to manufacture for someone else, a third party, or a competitor, and they think it might just be bad for their image. It's a strange situation because there are many food factories around—my former company owns and runs several hundred worldwide—which are not all completely used, not even close. On the other hand, if your branded company does not manufacture for a third party, someone else, a so-called co-manufacturer or co-packer will.

Food Industry R&D: A New Approach, First Edition. Helmut Traitler, Birgit Coleman and Adam Burbidge.

There are of course more fears to not wanting this than competition and bad image, such as loss of control and especially the potential loss of confidentiality.

Actually, the point that I want to make here is that irrespective of the definition and which type of food company we are looking at, at the center of it all is always a product that was manufactured by someone, either the own company or a third party.

8.1.1 Branded products or private label?

One could expand the definition of a food company to the typical representatives of the trade, the food sellers of any kind such as super-, hyper-, mini-, or mega food market (or any other name that you may find for the sellers). They are definitely a crucial part of the entire food industry but typically do not define themselves as food companies. There are, however, exceptions because all private label stores, such as Aldi or Migros, manufacture their products, either in affiliated manufacturing sites (almost factories of their own) or through third party co-manufacturers. So, it's really not so easy to draw a line between those in the food industry who purely manufacture such as Nestlé, Unilever, Kraft, Mondelez, Heinz, Cadbury, Mars, Pepsico, Coca Cola, and many others and those who purely sell. Because almost all of the latter, such as Walmart, Tesco, Safeway, Krogers, Giant, and again many more also sell private label themselves, which would qualify them for the club of the branded food companies because their private label is after all their own store brand.

I do admit that this becomes a bit confusing and this is on purpose. The setup of the food industry and a typical food company setup not only is difficult to define within the framework of the food industry at large but also from a purely inward-looking point of view, how a food company is typically organized and how its structure drives or is driven by the market needs, by consumers, by the trade, by stake- and shareholders, and all those who have a keen interest in a well-functioning industry, not least of all health and safety authorities in a global, a country-by-country, or a market-by-market view.

8.1.2 The food industry: A champion of complexity

Let me make it even more complicated: what about the agricultural industry, all of it? Every farmer, every agricultural association or cooperative, every crop manufacturer in the largest sense, potentially all seed and agricultural chemicals manufacturer is an important, if not pivotal, member of the food industry. After all, the slogan "from field to fork" is still a valid one, even if the pathways have become highly complex and even more highly regulated. And there is more: as a modern-day urban gardener, producing your own herbs, tomatoes, beans, peas, onions, or spinach, you are definitely part of the food growers and manufacturers. You are maybe not (yet) a member of the food industry because you use all raw materials fresh in your own kitchen. However, if you think of going into the business and get a food truck and go and sell your own homegrown products from the truck or from a stand at a farmer's market, you are part of the food industry.

Figure 8.1 illustrates the puzzle that represents the food industry.

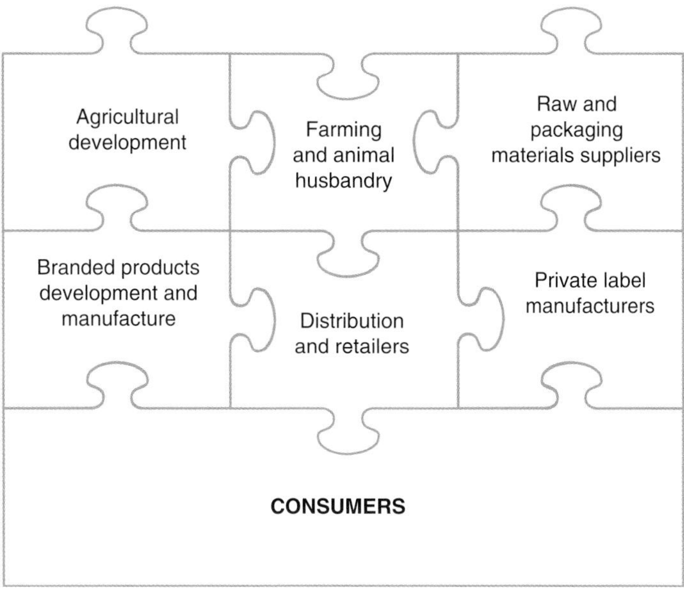

Figure 8.1 The food industry puzzle.

8.1.3 Some stories: Small food businesses and simplicity in their setup

Let me illustrate this by a personal story. My wife and I have really good friends in a lovely and small tourist town in the Blue Mountains, west of Sydney, Australia. They had been in the restaurant business most of their professional lives, which, by the way is yet another part of the so multifaceted food industry. After having sold their restaurant, they started a small food business on wheels, selling their own homegrown vegetables and other farm products as small ready-made dishes, almost like a fast food on wheels, at fairs and large garden parties. They definitely belong into my world; they are, by definition, a food company, entirely horizontally integrated. The large food companies believe that horizontal integration comes straight from the devil and must, by all means be avoided!

We have neighbors in California, who run a chocolate shop, which could be right out of the movie *Chocolat*: tasty and enigmatic hot chocolate beverages, mochas, truffles, and everything you may have dreamed of but never dared to taste because it modifies your body mass index in the hours that follow a visit to that shop. One could say "from heaven" without wanting to make this sound religious in any shape or form. They do not have a cocoa farm, so they buy cocoa and transform it to their finished products. Well, that does sound like the food industry: procuring agricultural raw materials, transforming them in smart and preferably proprietary ways, and selling them on to someone else who is in contact with the end consumer.

In our neighborhood, we have a wonderful coffee shop, which was the insider tip of the year 2012. In the meantime, as usual, it has become everyone's pick. The family who owns this in part has roots in Columbia and they have their own family coffee farm there, which supplies the coffee shop with a selection of high-quality beans. They have a little Probat® coffee bean roaster in the shop, and whenever they roast you think you might be in heaven, just better. This shop is a food or rather beverage company, totally. They have put the slogan "from field to fork" into practice, even if it translates to "from field to cup" in their particular case. They have not only stuck to making coffee beverages in the shop for immediate consumption or on-the-go, but they have also sold packs of roasted or roast-and-ground coffee to their clients. More recently, they have expanded to supply other businesses with their coffee beans, homegrown on a family farm in Columbia, transformed on a small German machine in their shop to a wonderful product; they have become true local beverage heroes.

8.1.4 How it all started

This, by the way, could be the story of any food company, growing from a family start-up, growing into a semi-industrial format and ultimately becoming so complex—procurement, raw material selection, price negotiations, some form of product development, manufacturing products, and distributing them to the initially selected or hoped-for consumer group or groups. Even the large food companies have grown from the first spark of recognition of a need, a problem, or an issue in the society or often a smaller group in the vicinity. The spark was translated into a possible solution, in the case of food, such solution is either a food product or a beverage of some sorts. To go beyond selling to your friends or neighbors the same day or the day after the product was made, one has to add a more sophisticated setup, a structure begins to crystallize, and the number of people and in turn complexity increase, first slowly and, if the initial success grows, so grow numbers of people involved and complexity.

Let me give a few historic examples of really old food companies that are still in existence and operation to this very day. I added two examples of really old companies outside the food industry, just to give you an idea what "old" really means:

Nakamura Shaji of Japan founded in 970, still operating
Marinelli Bell Foundry of Italy, founded in 1000
Weihenstephan Brewery of Germany founded in 1050
Acqua Panna mineral water of Italy founded in 1564
Salumeria Giusti of Modena, Italy, founded in 1605
Nestlé Company of Vevey, Switzerland, founded in 1866

There are many more, and more recent examples but this is just to illustrate that despite some of the parts of the food industry being of novelty (i.e. short-lived character), this industry has a real long-standing history behind and it is almost mind-boggling that one can sit at a table at the Weihenstephan brewery and almost 1000 years after the first beer was brewed there still drink beer of great quality, based on many hundreds of years of accumulated know-how. I use these examples that most of the examples of really old companies come from the world of food and beverages, and it can safely be said that any type of setup, structure, and organization in the food industry is based on many years of steady growth and slow evolution, always attempting to catch up with the present needs and, possibly daring to anticipate future needs. The industry

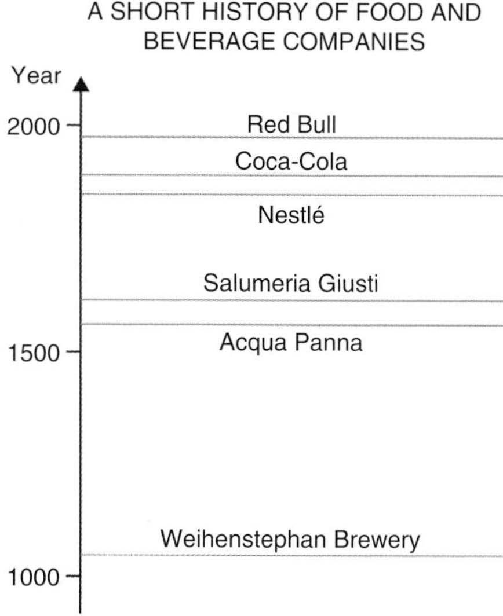

Figure 8.2 The big picture of the history of food and beverage companies.

will always publish slogans such as "getting prepared for the future" or "building the enterprise of the future" or "preparing for the challenges of the next decade" and probably a few more. In my personal experience, this is pretty much lip service and most of the time, remains lip service.

Figure 8.2 shows a simplified and selective overview of the history of some food and beverage companies. Many, if not most, companies were left out of this overview, not on purpose but to better demonstrate the historic big picture.

8.1.5 A bit of history: Strategic business units

Let me illustrate this at an example already earlier discussed in Chapter 2 and that dates back and was, even in those days, seen as rather revolutionary as far as a food company's organization goes. Twenty-five years ago companies such as Nestlé did not quite get visitors from outer space but from large consultancy firms such as McKinsey & Company. The big thing in those days was to restructure food and other companies by introducing strategic business units (SBUs). My former company was not the only place where this was introduced, but I happen to know this one pretty well. Interestingly enough, SBUs were seen as the solution to structural complexities, especially in large companies, and were seen to help streamline the business organization and increase efficiency by centralizing certain organizational and managerial functions for the business in question under one head and one roof.

In the beginning and especially taking the then existing complexity into account, this was seen by many as a good move forward. In those days, the Nestlé Company was organized by some kind of central business structure relative to the many product groups. The business was assisted by a rather complex technical support and R&D structure, the latter again centrally managed yet geographically spread all over the world. On top of this came an organization that had semi-operational responsibilities, which spanned over what was (and still is) called "Zones."

There were five at the time; there are still three today. These Zones were simply organizational conglomerates of geographic areas, located in the headquarters of the company. At its simplest it was something like "the Americas," "Europe" (in the meantime including Middle East and some other regions), and "Asia, Oceania, Africa," which has also undergone slight modifications in the more recent past.

8.1.6 It's getting really confusing now

But wait, that's not the end yet; there is more (almost sounds like QVC teleshopping): it was recognized that every single product has not only to be manufactured in a particular place but also should be sold more or less close by, necessarily in a country, which in this structure is called "market." The markets have certain responsibilities, the ultimate one being P&L. However, the SBUs have some responsibility too, which was at a later stage defined as "accountability," and we were all asked at a certain point in time to learn by heart the distinction between "responsibility" and "accountability." If your head started spinning a while ago, rest assured that it might continue to spin for awhile to come because this is not the end of the story yet and there are more structural and organizational twists. The first one is the distinction, if not even, "demarcation" between the specific functions of any given SBU (reminder, if it wasn't clear before: SBUs are product group specific) and the R&D groups on the other hand. In this particular event, each SBU had to negotiate with each corresponding main R&D center (the one that was most active in the product group in question) and hash out the answers to the questions: who is doing what and who is responsible for what part of the various activities. On a side note, many R&D centers did (and still do) cover more than one product group and in turn, one product group could easily be covered by more than one R&D center. Many documents were created amply showing the strength and usefulness of word processors, mainly written in two columns across each page, one column for the SBU, the other one for R&D. Funny enough, such documents were fondly called "Repartition" documents, lovingly using a term that was created in a language called "Franglais" and was especially spoken by our British colleagues trying to blend in a French-speaking environment.

8.1.7 One important change of R&D setup as a consequence of a changing business structure

The documents, whatever they were called, had one really excellent effect: at least on paper, everyone knew what he or she had to do and what was expected from each group. An important additional consequence was the creation of SBU/product group–relevant R&D centers, which had to be given a different name now because the old one (R&D center) apparently was not descriptive enough for a center that had all necessary elements from recognition and translation of consumer insight to basic technical, and as needed, scientific competence development, to product-specific new product development, technical assistance, and required engineering competence all under one roof and was therefore called Product Technology Center or PTC.

I promised you another twist to the structure that was introduced since the inception of the SBU structure, which was mainly the result that several important new business acquisitions happened and that were let to operate in more independent, stand-alone fashions and were thus set up as such. However, while the SBUs were located centrally in the head office, such stand-alone businesses had and still have their seats somewhere else, mostly in the largest and most important market. The main unifying element between SBUs and stand-alone businesses is that they both draw from R&D resources, which are still to some degree centrally managed, although their programs and work directions are mostly influenced by the different SBUs and businesses. I do get it: your head is full spinning and you may have lost the red thread completely. Let me bring you back: the major influence was that the food industry was subject to organizational and content changes through the restructuring and introduction of new setups.

8.1.8 What's first: The chicken or the egg?

Fast-forward and as an important intermediary conclusion and as if it was not obvious anyway: R&D only changes if the company changes. The big question however is, what comes first: the chicken (the business) or the egg (R&D)? The other important question that needs analysis and some profound discussion is whether the majority of successful food companies (and by definition they must be successful otherwise they would have disappeared a long time ago) is successful because or despite the current setup and especially the role and function that their R&D plays today. To find the right answer to "because" or "despite" is not an easy undertaking and probably depends on who gives the answer and who defends either the one or the other position.

Following the "because" faction would suggest that everything is in order and we may continue as in the past. Following, however the "despite" supporters would probably mean that the R&D group is to some degree irrelevant, or to say it nicer, less important to the success of the company and only exists because management, which believes that the "R&D organization costs twice of what it should actually cost but we don't know which half to cut," does not dare to change the situation in dramatic and maybe catastrophic ways. I truly believe that the way out of this dilemma is not just eliminating half of the resources but reorganizing and totally reorienting its competence base, especially the content of the R&D group's work, pretty much in line with what was already briefly discussed in Chapter 6, the consumer perspectives for change of the R&D approach in the food industry.

Let me close this paragraph by saying that the entire previous discussion is mostly relevant to large food companies and hardly applies to smaller, possibly family-owned SMEs. Neither do they have money and time nor the need to manage their fairly low degree of complexity with the means and setups that are typical for the large corporations. This is not to say that they have an easy life and none of it applies to them, but the solutions are typically much simpler, more direct, and based on personal interaction because most of the players in SMEs happen to be in close-by locations, thus making direct personal contacts much easier and more frequently.

Figure 8.3 depicts an attempt to paint a picture of "The Complex Company." It is a mix of different approaches found in different companies but in the large companies this picture is frighteningly close to reality.

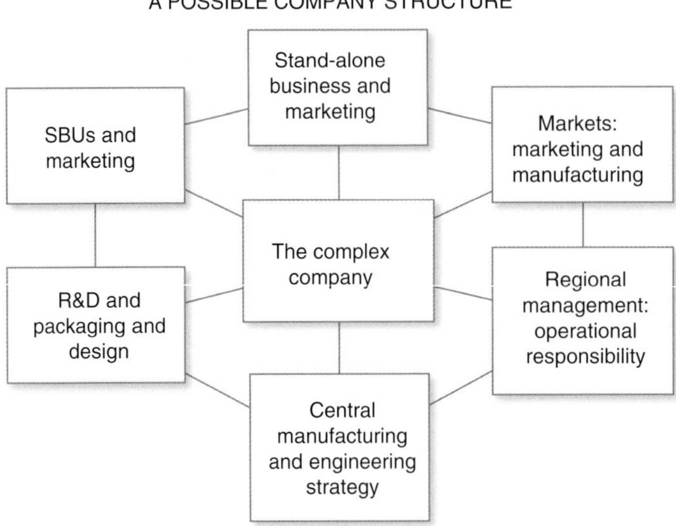

Figure 8.3 The really complex company.

8.2 THE FOOD INDUSTRY: AN EASY MONEY-MAKER OR A DAILY BATTLE?

Nothing comes easy, especially not in business. So, the simple question is, how does the food industry make money? The answer appears to be trivial: by selling food and beverage products. At least that's the traditional business model, the one that was in place when Weihenstephan sold its first beer to consumers around 1000 years ago and that is still valid today for most of the food and beverage companies selling Nescafé or Coca Cola. Clever marketing managers will, however, emphasize that they are selling lifestyle, a dream, the strong belief to belong to a trendy group of consumers, or simply for comfort and convenience. They sell the belief that the product does something good for me, from a simple "quenching my thirst" to "rejuvenating and invigorating." It's of course not different from other industries and other products, however, the approach is old, not to say at least 1000 years.

8.2.1 Marketing is really old, really, really old

The monks who brewed and sold their beer to travelers, tradesmen and people from their neighborhood sold relaxation combined with reinvigoration and quenching thirst. It was and is to this day the sale of a lifestyle, namely to rest in a beer garden, possibly under then already-old chestnut trees and drinking a more or less cold beer fresh from the cooler cellars. And these monks and their successors of present day made money with what they did, otherwise they would have given up. And despite beer brewing being something traditional, they felt the desire and the need to always improve and develop something new. Not so much maybe in the art of brewing but certainly as far as ingredients optimization as well as surrounding technologies was involved.

One can argue that it's never a good idea to change a winning team or the change an approach that has worked and still does. However, it is my firm belief that it's overdue to change because the food industry cannot continue to make money the way it has since pretty much 1000 years ago. "Something's gotta give!" And yet the industry does not seem to be ready for change, at least not the kind of change that is almost visible as the proverbial writing on the wall.

Despite the fact that it's certainly not an easy undertaking to sell food and beverage products consistently successfully, year in, year out, and into the future, it has, after all been a fairly easy ride for the industry despite many hardships mostly through self-inflicted glitches. There is so much at stake, and if you take the large food companies as an example there are dozens of billions of dollars or euros or Swiss francs at stake every year and hundreds of thousands employees worldwide. So, it's not that easy at all. Or take for instance smaller companies, family-owned, SMEs or even below that margin, the food startups, which can be found more and more frequently today. Much is at stake, be it the consumer satisfaction leading to repeat purchase or stake- and shareholder satisfaction, investors' satisfaction, government authorities' satisfaction, consumer groups' satisfaction. It's become complex; although the direct sales of my urban garden-grown produce simplify this entire framework a lot.

My call is for change in many ways and I have briefly insinuated this in previous chapters. The food industry is increasingly criticized for the core that it represents: industrial food. Companies can hide the facts as much as they want by manufacturing so-called *semi-finished products*, suggesting that the person that uses such "modules" add her own personal touch during the preparation and thereby gets the feeling that he or she has contributed to the final dish. Fact is that we are still dealing with an industrial product of sorts, and industrial food is increasingly at the center of critique by many groups, most prominently consumer and other pressure groups. The points mostly criticized and debated are not only that such food has been manufactured in big portions in factories, detached from the consumers' homes and kitchens, but even more importantly, it's about the sustainability of using agricultural resources and that it may not be the best way to do things. I do realize and have to admit that the food industry has undertaken big efforts to integrate sustainability right into the center of every new food product development, manufacture, marketing, and sales and has to be complimented for this. However, these efforts have not always been recognized and acknowledged by consumers, and much mistrust can still be found. The food industry could either do more in this field or turn their attention to the future and develop new business models, which make them less dependent on the necessity to defend their well doing even more but open totally new directions that may be unthinkable today but may be the talk of the town tomorrow.

8.2.2 Can the food industry turn to a new direction and new business model? Is a revolution possible?

Let me expand on this topic. The topic of this chapter is "the industry perspectives for change to the R&D in the food industry," and this is at the heart of this analysis and discussion. If the industry heads toward a new business model, it is clear that R&D has to change as well. I have asked the infamous question of what was first, the chicken or the egg? There is, of course, no real answer to this and the pragmatic answer has to be: whatever comes first. Let's look at the two possible scenarios for whatever reason, the company decides to change its business model, either out of necessity (for instance P&G back in 1999–2000) or out of foresight (no real example available, unfortunately) and therefore reorganizes and redirects its R&D division. Or the R&D division and some of its visionary leaders and doers came up with outstanding results potentially paving the way toward new opportunities, and the company adapts or even rather dramatically reinvents its business model. The latter is rather unheard of in the food industry.

When Dietrich Mateschitz concocted Red Bull® pretty much in a bar in Phuket (Thailand) in the early 1980s and launched the product first in Austria in 1987, it was such a finding, an inspiration, an idea, or just a great conviction that shaped the business plan of the company in every way: development, manufacture, target consumer groups (initially snow boarders), distribution, advertising (comic spots, "gives you wings"), and sales, and even pricing. It was and still is pretty daring to sell such a beverage for close to three times the price of Coca Cola® or other soda drinks. And yet it worked because it was so daring at the time. In the meantime, their business model has changed quite dramatically and much of the notoriety that Red Bull has garnered and still does stems from areas totally outside the food and beverage industry such as motor sports (Formula 1), flight shows and competitions, soccer and many other sports, and entertainment-related areas. Red Bull® truly sells lifestyle much more probably than any of its competitors and could easily be considered to be a member of the entertainment or sports industry rather than the food and beverage industry.

So how did this influence R&D? Well, there was an extremely important learning: much or most of the science was available and well known, such as for instance the stimulating effect of the ingredient Taurine. Caffeine is well known, if not to say widely known for its stimulating effect and was therefore a natural sibling in the beverage. Most R&D was based on consulting available nutritional and medical literature and doing homework regarding legal status of the various ingredients in the different markets where the product was going to be sold. This was, in my eyes not flawlessly done, because it was known that in those days and for quite some time to come, that Taurine was not permitted for instance in France. It took 20 years until 2008 for Red Bull® with Taurine to be sold in France. Therefore, most of the R&D was actually "D" both for formulation as well as packaging and design of the beverage can and could therefore be performed rather expediently and with limited resources as is typical for a startup company, which Red Bull® definitely was in the mid 1980s.

8.2.3 Let's do this together

There is important example for a third way: neither business nor R&D are the sole drivers for a new business model but both partners have a say in where the trip will be heading to and how it may be achieved. However, this approach is most likely not working in most established food or beverage companies, irrespective of their size. And yes, it becomes even more difficult if not close to impossible to initiate such change in large food companies. And yet, change has to come, especially because of the increasing pressures from the end consumer and consumer groups. And there is more writing on the wall when we look at the industry of bottled water. Perception is everything, even if it goes against reality. From historic examples we know that rupture points, important almost-catastrophic events, are the moments when change happens because it has to happen, there is no other alternative. I already mentioned the example of P&G. There are actually fewer examples that I could find that have led to a successful rising from the ashes than a one-way downfall toward oblivion. The "late" Union Carbide is such an example that never recovered from a chemistry accident on December 3, 1984, in Bhopal, India. This event literally brought down the entire company, and in early 2001 Union Carbide was ultimately acquired by Dow Chemical®.

This is of course not the outcome that anyone wishes from such rupture point events, and it is therefore useful to constantly scan even for the slightest bumps in the ongoing history of the company and be prepared to reply to this in the most successful ways, at the same time using such moment to induce change. In using the word *change* quite a few times already and most certainly as a recurring theme in the entire book, let me say something possibly controversial: I truly believe that change for the sake of change is actually a good thing, contrary to the common belief that change should only happen when there is no other way out from a present problem, downfall, or real rupture point. I am convinced that change is properly rehearsed by changing one's ways ever so often and looking for new approaches to old or new issues and opportunities constantly. Desire to change the approach as well as content of one's work has to almost become part of the DNA of every group in the company and that's why one does not have to wait for a catastrophe to happen but can ride on a wave of anticipation.

8.2.4 Easy money or daily struggle?

Such desire, because it is in the DNA of everyone in the company, is now driving constant, if not perpetual, change all the time and in every corner of the company and the question of who drives change becomes obsolete. The real question, however, is change also driving the company forward to become more creative and innovative and bringing about real new and surprising products and services for everyone on the receiving end of industrial food and beverage products? I shall analyze and discuss this in some detail. But before I go there, let me attempt to answer, is the food industry, a food company, an easy money-maker or does it fight daily battles to succeed in a large and competitive marketplace with relatively low margins at least as compared to other industries. Food and beverage in almost all its forms is considered a commodity, at least most of the products manufactured, and therefore is confronted to such low margins quite naturally.

That is not to say that food and beverages cannot make big, sometimes seemingly easy money and some flagship products in the industry such as soluble coffee, all kinds of chocolate and candy products, as well as old and traditional brands in the beverage, baby, and infant segments are real money-makers. Some of these products have not undergone any major change for many years, yet still are given great attention and support by both the companies, who own these products, and the consumers, who love them. One could think that this is just a situation commonly called *milking* and easy money is made from these products and brands. Yes, but it still is a daily battle to succeed every minute or even every second somewhere in the world to sell one or more of these products to consumers. I tend to think—and maybe I am wrong—that the situation is probably easier for smaller food companies, which can focus on fewer brands and a smaller and more defined environment to sell these brands. After all, the simple answer is: it's both, easy money and daily struggle!

8.3 IS THE FOOD INDUSTRY REALLY INNOVATION DRIVEN?

This question is an often asked and debated one. Most of the time, the answer is rather in the negative, however, I would see this much more differentiated. I have written about this in *Food Industry Design, Technology and Innovation* (Traitler, Coleman, & Hofmann 2014) and have also briefly covered this topic in previous sections of this book, especially the topic of innovation versus renovation in this industry. The first question that may come to mind in this context is: can the food industry bring real innovation to the world, especially of course to the consumers? Let me analyze this from a historic perspective and let me quote the following list:

"Enchanted Learning" lists the following food-related inventors and inventions:

- Bar Code, also called Universal Product Code or UPC, which was first developed and patented in the 1950s by inventors N. Joe Woodland, Bernard Silver, and David J. Collins.
- Bread Slicer by Otto F. Rohwedder and Gustav Papendick in 1928.
- Burbank potato (or Russet Burbank) bred by Luther Burbank in 1871.
- Can and can opener invented in 1810 by Peter Durand.
- Carbonated water first produced by Joseph Priestly in 1767; first carbonation device invented by Torbern Bergman in 1770.
- Chocolate chips invented by Ruth Wakefield in 1930.
- Coca Cola first invented by John S. Pemberton in 1886.
- Cotton candy soft confection invented by William Morrison and John C. Wharton in 1897.
- Dutch chocolated (rather "dutched" or alkalized chocolate) was invented by Coenraad J. Van Houten in 1828.
- Hot Dogs first sold by German immigrants in New York City in the 1860s.
- Kool Aid (powdered drink) invented by Edwin Perkins in 1927.
- Life Savers peppermint candies invented by Clarence Crane in 1912.
- Marshmallow was already made by ancient Egyptians more than 3000 years ago.
- Mayonnaise probably invented in 1756 by a French chef working for the Duke de Richelieu.
- Mechanical reaper for improved harvesting invented by Cyrus Hall McCormick in the mid-1800s.
- Microwave oven invented by Percy LeBaron Spencer first commercialized in 1954.
- Pasteurization invented by Louis Pasteur in the mid-1800s.
- Popsicle invented by Frank Epperson (at the age of 11!) in 1905.
- Peanut based products first developed by George W. Carver around the turn of the nineteenth to twentieth centuries.
- Potato chips invented by George Crum in 1853.
- Refrigerator had several periods of invention: first in 1748 by William Cullen, followed by Michael Faraday in the 1800s, Oliver Evans in 1805, John Gorrie in 1844. First electrical refrigerator invented by Thomas Moore in 1803; first commercial refrigerator sold by the General Electric Company in 1911.
- Sandwich invented by John Montagu, the 4th Earl of Sandwich around 1762.
- Sugar processing patented in 1864.
- Teabag invented by Thomas Sullivan around 1908.

8.3.1 Innovation in the food industry is rather an antique affair

This list, as incomplete it may be, shows mainly two important points: the first one being that most food industry innovations seem to be tool- and device-centric and secondly, the most recent innovations on this list date back to 1950s! I do know for a fact that there were quite a few real innovations coming from the food industry since then, hypoallergenic infant formula just being one of them. I do not want to stir any controversy in the area of infant formula, but it is a fact that partial protease induced breakdown of difficult to digest dairy proteins to smaller peptide molecules has helped a large number of babies who had received formula after several months of having initially been breast fed. Given the apparent scarcity of real innovation and breakthrough in the food industry, it may safely be assumed that the food industry is not really innovative and relies much—too much?—on what is typically called *renovation*. Renovation is mainly face-lifting of products or packages, whereas innovation is different. It is surprising, making sense in the eyes of the consumers, is cool, is new and better than before, is useful, funny, and entertaining, a great solution, is progress, really new, unexpected, and retrievable and reusable. Innovation is some or most of this list and maybe more, something that you would add and which I have forgotten to mention.

Innovation is my preferred topic, as much as it may be overly used if not abused in many discussions, debates, business plans, newspaper articles, and business and development plans; and in my own deliberations and workshops that I hold on the topic. On the other hand, if one does not talk about innovation and always bring this topic to the front of the debate, nothing will happen. Even constantly talking about it does not bring the desired results, as can be seen in many food companies. I repeat what I have said a few times already: the food industry may be great in renovating products and services but is notoriously incapable of real innovation. That needs to be changed and that is the real push that the industry—every company—needs to exert to its R&D organization: change your ways and innovate. And yes, there was a minor additional aspect to this: we, the business, are ready to accept real innovation and will implement it, even if it may initially cost more. There is no real innovation at zero costs, although this is the real secret or rather not so secret dream of every business responsible in the food industry.

8.3.2 IBM or Kodak: Which would you rather follow?

If this change will not happen soon, many food companies, especially the large ones, will have their "IBM moment," and if they still will not react accordingly, this moment may turn into what can be called the "Kodak moment." Not wanting to denigrate either of these two prestigious companies, their examples and how they reacted to the writing on the wall in their respective industry fields stands for the partial or total unwillingness to read them and take them for what they stood for: a cry for radical change and they did not listen as they should have. IBM certainly reacted slightly better than Kodak and that is reflected in the present-day situations of these two companies. Don't get me wrong, they were and are not the only ones who were unwilling to change in their business approaches and business models. They eventually did change but for some, it was almost or altogether too late.

The Economist (2012) wrote about Kodak.

> *Could Kodak have avoided its current misfortunes? Some say it could have become the equivalent of "Intel Inside" for the smartphone camera—a brand that consumers trust. But Canon and Sony were better placed to achieve that, given their superior intellectual property, and neither has succeeded in doing so.*
>
> *Unlike people, companies can in theory live forever. But most die young, because the corporate world, unlike society at large, is a fight to the death. Fujifilm has mastered new tactics and survived. Film went from 60% of its profits in 2000 to basically nothing, yet it found new sources of revenue. Kodak, along with many a great company before it, appears simply to have run its course. After 132 years it is poised, like an old photo, to fade away.*

The case of IBM is not only a bit older but, despite having similar writing on the wall, but it turned out differently. It was also more of a wakeup call for IBM, which happened in the mid-1970s with the onset of the personal computer revolution, which IBM did not want to see at first and when it ultimately jumped on the train it was too late. It is of course easy to write this in hindsight. As of the mid-1990s IBM took action by modifying their business model: they shed the (for them) unprofitable PC hardware business, they turned again to a professional mainframe server business, and they added a global service business that became a leading technology integrator. IBM did of course a lot more, and it would go beyond the scope of this discussion were I to mention all their feats and achievements of the last 20 or so years.

8.3.3 Change or perish!

The main difference between the two different companies is of course that the one, Kodak, had a really big downfall because change came unwilling or not at all and the other, IBM, was willing to change and adapt. There was one more big if not decisive difference; however: Kodak in the mid-1990s was at the height of success and the waters around the company appeared to be calm, too calm. IBM on the other hand was in difficulties, already in the mid-1970s and then again in the mid-1990s; they were used to trouble!

Should companies therefore look to make trouble for themselves so that constant change and adaptation becomes a natural process within every company and everyone in the company is not only on constant lookout for trouble and opportunity but is also listened to when he or she proposes change? I truly believe that this should be a way of operation for every company and especially every food company. Disruption, turmoil, and trouble should be keywords and descriptors of situations that everyone in the company can and will fully embrace. And that's the kind of industry perspective that should influence and steer the work and direction of the R&D group in every company. The danger of complacency and deliberately looking past opportunities exponentially increases with the size of the company.

8.3.4 Small is beautiful and creative

I started this section with the question whether the food industry is really innovative and have already repeated a few times that I strongly believe that the food industry is not. First of all, I have seen time and again that the rate of success of innovative products, processes, and services is higher in smaller food companies and diminishes fast with increasing size of the company. Just check for yourself, for instance in the area of new packaging development. Visit different packaging suppliers and converters, different most of all in the size of their business. Visit their product showrooms and you will find dozens if not more creative and innovative packaging concepts in the smaller companies (less than $100 million or even $50 million annual revenue), whereas the bigger companies (let's randomly say above $300 million annual revenue) will mostly tell you about their wonderful cost-saving stock, mass-produced items, which only excite the finance officer of the food company but certainly not the consumer.

Why is that so, why has this been like so for quite some time already? The answer is simple: large companies typically sell cost reductions or the results of such cost reductions, whereas the smaller ones fight for survival by selling the surprising, the creative, and innovative solution that the big ones have no time (or are unwilling) to pursue and sell. I admit that I paint with a rather broad brush, but just look into your own company or companies that you know about or have read about. I am linking this analysis and discussion to the food industry and do realize that this may be quite different in other industries. I am adamant: large food companies are not innovative and you can check for yourself by checking out supermarket shelves and the types of products that you may find there coming from different food companies. I really encourage you to do this assessment and analysis yourself and write back to the food company of your choice—target that rather is—and ask them why they do not have more exciting and innovative products and packages in the marketplace. Some companies may even write you back, although I wouldn't hold my breath.

8.3.5 Change your business model

This industry, especially its large members, need to change and they need to change soon if they don't want to fall back or even disappear into oblivion. Such change, especially demanded by consumers and consumer-related organizations, has to come and will have to cover all parts of the value chain. Innovation is still the name of the game and the change has to be radical. There has to be disruptive innovation in the area of agriculture, procurement, and supply chain, and there has to be radical innovation in the way that the industry brings food and know-how around food and beverages to the trade and ultimately to consumers, including new ways of manufacturing, especially making industrial food again attractive to consumers. This is quite a handful and is not an easy feat.

In my eyes, however, such change is not only necessary but is only the beginning of much more radical change that potentially looms around the corner. I strongly believe that there is writing on the wall that will force food companies such as Nestlé, Kraft, Mars, Unilever, Cadbury, Pepsi Co, and others to radically change their business model and go away from a business to consumer (B2C) to a business to business (B2B) approach, fully embracing the trade as the ultimate customer and vehicle to the end user, the consumer. You may argue that this is the case already today, however, it is not. Every action, every project, and everything that is done in a food company today, is done under the aspect of "consumer benefit," all "consumer-insight driven." This is a strange situation because quite obvious already, the trade is the real customer, a customer that has been seen as adversary, even enemy of food companies not so long ago. It would be so easy to embrace this situation to the fullest and act accordingly.

What food companies in the future will have to sell is not so much a consumer-insight–based food or beverage product but real knowledge when it comes to good food, health, nutrition, and lifestyle. And lifestyle is something that some food companies know very well. I have mentioned the case of Red Bull before; this is a company that sells lifestyle more than anything else. The canned beverage is almost an accessory and a side thought to the mainstream: lifestyle and dreams. I am not suggesting that every food company should do alike but needs to do differently from what is done today. Ultimately, this will influence the ways that any R&D group in the food industry conducts its work and how it will have to revisit content and especially desired outcome of their work.

Such a dramatic change of the basic business model entails an equally dramatic change of strategy and tactical action points for food companies considering such new directions. The major target point for a new strategy is service and using knowledge as sellable good. Food companies will have to become real service companies more than anything else selling new types of products such as know-how, nutritional understanding, good and beneficial nutrition, food and health, and many more. Know-how is the new bitcoin and service is the new mantra. This will create a major uplift for the importance and role of every R&D group in every food company, irrespective of its size.

8.4 THE PERCEIVED VALUE OF THE R&D ORGANIZATION: IT'S IN THE EYE OF THE BEHOLDER

Real value can hardly ever be measured correctly, let alone get agreement of everyone involved as to how much a product, a service, a patent, or know-how is worth and to whom and in which context. I admit, I made it a bit complex on purpose, but the truth is that value is most often in the eye of the beholder, especially when we discuss *perceived value*. Depending on whom you talk to in a company and especially in a food company, about perceived value of the R&D department you will get all kinds of different answers. One line of thought you often will hear is that R&D is highly overrated and typically underperforms, that it is too big as an organization and should be redimensioned, and that it's not enough under the control of the business. Another perception that is frequently expressed is to say that R&D in principle is a good thing, but it could entirely be out-sourced and there is no need for an R&D group inside the company. The business will deal with the results appropriately and only ask for external R&D support on an as-needed basis.

8.4.1 Why R&D is useless...

While the former perception (too big and not responsive enough) is often expressed by managers, who have a deep mistrust of every department in the company except their own, the latter one (all out-sourced) is mostly heard from the "kingdom-builders." They want to be in control and by proposing the dismantling of the internal R&D group and dealing directly with paid for and on the go external R&D they believe to be able to control everything and become the real kings of their respective businesses. There are of course many other perceptions of the value of R&D in the eyes of the business side of the company, which can be situated in between the two positions.

The main points of criticism directed toward R&D in the food industry are almost always the same and can be summarized as follows (and I have mentioned them already):

- Too expensive
- The wrong results and output
- Too academic
- Or, alternatively, not enough academic (depending on the CTO's preference)
- Hiring the wrong resources and expertise, the wrong science and technology for the required needs
- Too slow (actually that most of the time is "toooooooo" slow!); no sense of urgency
- Not responding to the real needs of the company
- Too far detached from the business reality (the one defined by the business)
- Cannot be trusted because "they do what they want"
- Ultimately, does not understand the business

8.4.2 And why R&D is great!

You may find more points of content with regards to R&D in the food industry and I encourage you to list them all because it is an important part of the "therapy" that I suggest in this section, which should lead to emphasizing rather the importance and reasons to be of any R&D group inside as well as outside the company. So, let me find the most valid, plausible, and logical arguments why an R&D organization in any food company is most valuable and much needed:

- Creates and develops innovation and innovative solutions
- Has expertise that supports the business in important ways
- Is the go-to when it comes to trial and error of concepts
- Adds credibility to the company at large
- Adds value by simply acting as fortress for know-how and IP
- Gives answers to questions relating to nutrition and health
- Supports manufacturing in important ways by giving much-needed technical assistance
- Represents the best platform for learning and creating an incredible talent pool for the company
- Is required to work at eye level with external expert resources
- Helps to alleviate the tax burden of every company and for whichever legal/financial setup

Again, this is definitely not a complete list and I encourage you to add as many arguments as you see fit and that support the list of reasons to be for an R&D organization inside your company. The real question of course is, how much not only perceived but also real value can the R&D organization bring to the company in a consistent fashion? And this is quite a different caliber of argumentation than just listing the real or perceived positive or negative points. This can become quite a numbers game and in most instances is not won by the R&D, hence the constant pressure by the business on the R&D organization in the company. Even if what I call numbers game appears to be a rational undertaking and not suspected of being partial, it is however highly irrational and outright emotional. Why do I say this? Well, I have experienced a great number of instances when R&D's added value to the company was discussed and numbers were used and misused, interpreted and misinterpreted, and conclusions were reached, which at the end of the day were the ones that at the beginning had already been decided on: a lot of fog and chimera for a predictable outcome.

8.4.3 It's because of the tax man

And then the next round of cuts and downsizing can begin. I do remember a discussion that I had with one of my former bosses when we debated the question of how to make R&D really valuable to the company. My short reply was simply this: we are losing time because if the company were to dismantle its internal R&D organization it would lose more money to tax paying than the entire R&D spent. It may sound like a weak and cynical argument, but it was simply the truth: the tax advantages gained through the sheer existence of the company's R&D group outweighed the investments into the same R&D group and its activities.

I had of course no real insight into the exact financials of the tax write-offs through R&D, but I knew pretty well how much the company spent into the activities of its R&D. Although many like to call this spend *costs*, I prefer the term *investment*, and an investment it is, I do insist!

8.4.4 The sense of urgency is really missing

This is the moment when you realize that in addition to the sheer numbers (clear and rational), the element of emotions and irrationality come to the fore and almost take center stage in the discussion. And there is no one winner, even if the business gets its way and cuts yet another portion out of the R&D budget. In the long run it loses; it's like ordering austerity measures to a country, knowing exactly that no growth will be generated through such measures. This, however, is not to say that there is no such thing as a healthy cutting down on resources and activities, provided it is done in a measured attempt to increase creativity and the sense of urgency in the company's R&D organization. And this is probably the most crucial and most critical argument that unfortunately speaks against the folks in R&D: the missing sense of urgency and being too slow! This is probably the most frustrating element of all for the business side of the company when asking the R&D side to bring solutions that fit, that are relevant, more or less affordable, and do not take a decade to come up with.

This slowness and this missing sense of urgency is the biggest perceived value destroyer in the eyes of many and many even in the R&D organization itself. So why does no one do anything about it? I have no real answer to this, and it has always been an enigma for me, even when I was part of such R&D groups and was equally guilty of not having acted enough on this enormous downfall, yet tried I have! So how can the super slow "Moloch," called R&D, be

WHY R&D IS BAD	WHY R&D IS GREAT
Too expensive	Greater innovation
The wrong results and wrong output	Has expertise to support business
Too academic	Can do fast trial-and-error approach
Not enough academic and scientific	Adds credibility to company at large
Wrong resources and expertise hired	Adds value through creation of IP
Far too slow and no sense of urgency	Has nutrition and health-related expertise
Not responding to real needs	Supports manufacturing and gives technical assistance
Too far detached from business reality	Represents talent pool for the company
Cannot be trusted, "they do what they want"	Works at eye level with external expert resources
Does not understand the business	Helps alleviate tax burden for the company

Figure 8.4 Two opposing views on R&D.

tamed and brought to run fast instead of standing still? And run fast it must, or else: further cuts, downsizing, let go people, downward spirals, and the rest may be history.

There are many examples of such downward spirals in the more recent history of companies, luckily for the food industry more outside its boundaries than within. I do realize that there are occasions at which cutting back and saving money is the only way out. My criticism, however goes to the group of managers that have let the situation slip to such an extent that cutting back becomes the only viable way out, yet not the right way forward.

Figure 8.4 lists the most often heard arguments of two opposing views of R&D: the bad and the great.

8.4.5 "Good-weather" versus "bad-weather" managers

Why do these situations happen at all and why do companies have to downsize and cut back? There are many facets to this question, and unfortunately as many answers. This is of course a bad thing because many answers represent cause for many excuses, and everything can be explained.

If I had to choose one answer, then it would be that it is typical for the food industry and does not necessarily reflect on other industries. Most top managers in the food industry are "good-weather" managers. In other words, they shine when everything goes well, and they panic when the times are tough. It's the opposite of "the tough get going when the going gets tough." I have seen this so many times and was always astonished how this could happen. I have to add that luckily I had experienced many good years with my former company and have seen many good years in other food companies too.

So the struggles were rather "fair-weather" ones; however, I begun to observe more "bad-weather" struggles more recently in many food companies. It appears that this is a situation that the present-day management in many food companies cannot properly master, and they revert to cost cuts, budget cuts, and an increasing number of constraints. I do not say this lightly but it's the same situation that a soccer goalie faces who has not been tested during 80 minutes of the match because his team was so superior. Then comes the first serious wave of attack from the adversary and boom, he gets a goal! Same thing here: good-weather managers had a superior team around and under them and did not see and feel the necessity to train for the bad times that might come.

8.4.6 Constraint is good, smartly dealing with it is better

In previous chapters I have discussed and analyzed the value and positive impact that constraints can have on the performance and efficiency of everyone in the work environment. However, I have also emphasized the need for measured, sensible, and not just symmetrical cutbacks if management does not want to destroy the morale and motivation level of their employees. More is destroyed than gained through badly communicated and broad-brushed symmetrical cutbacks. This is when a bad-weather manager can shine and pull his or her people forward to new heights despite more or less deep cuts into resources, content, and direction. This is especially true for every R&D organization, which has its eyes set on more long-term goals including the hiring of appropriate resources with specific backgrounds that cannot just be redirected and re-educated over night. This is the real industry perspective for change, especially in the R&D group that I can carve out and formulate as the major conclusion for this section if not the entire chapter: Think long term and anticipate future changes even when they are seemingly far out because come they will. Putting your head in the sand will not make a coming change go away.

Food industry managers at any levels are masters in denial, especially when it comes to "talking away" possible future threats by piling up past successes as proof for a successful company and everyone believes them. Even the more mainstream financial analysts do not always look under this pile because they are blinded by the success stories. More self-critique is absolutely needed and the courage to discuss changes, especially potential radical changes to the company, which might first be addressed by the company's R&D organization. It is not helpful to brush aside every attempt to recognize and discuss a future that may be different from the present and think that it is easier and safer to just remain and wait out, at least for the moment. So let me close this chapter by saying that change is in the air, reacting to the new realities is a necessity, and it shall ideally start in the R&D organization by challenging all elements: structure, content, people, expected outcomes, timelines, budgets, dependence versus (partial) independence, and a few more aspects that I shall propose, analyze and discuss in Chapter 10.

8.5 SUMMARY AND MAJOR LEARNING

- So-called "branded" food companies develop and manufacture their own branded products and shy away from manufacturing private label products for those large distributors who mainly sell their own store brands. This led and still leads to a rather unhealthy situation, putting the two players, manufacturer and seller, respectively, on opposite, competing sides.
- There is no clear distinction between food companies, which exclusively manufacture products or deliver services and those which are mainly distributors, as the latter often entirely or partly sell products, which they manufacture themselves or have them co-manufactured in affiliated manufacturing sites.
- Traditionally food companies develop, trial, manufacture, package, market, and sell food products as well as services. Although all food companies pretend that they do all this for the consumer, one has to realize that they do all this for the trade, which is their real customer.
- The agricultural industry and agriculture development at large are important members of the larger food industry. The slogan "from field to fork" is more valid than ever and describes the interconnectedness of all members of the food industry.
- Large companies in every industry have a strong tendency toward increasing complexity of every step of the value chain and the business overall. One could almost accuse large companies of having a love affair with complexity. On the other end of the spectrum, small companies cannot afford such complexity and can therefore act and decide faster and more efficiently. A lesson that the large corporations should learn from the small players.
- Food and beverage companies are typically old, really old and the oldest one is approximately 1000 years old and still operating strong. Complexity and quirky business habits have therefore had a long time to grow and infiltrate entire companies.
- Companies have always looked out for new organizational structures, all with the goal of reducing complexity and better control of the business. The most dramatic change came with the introduction of the so called *strategic business units* in the early to mid-1990s. Although the intentions may have been good, this definitely led to increased confusion and more complex intercompany dealings. R&D organizations were affected, and although things should have become easier, they actually became much more complicated, by adding an additional intermediate between the business and the R&D group.
- The question of whether the food industry is an easy money-maker or has daily battles to fight was analyzed and discussed and the simple answer is both.
- Another important question was raised: can the food industry actually be really innovative and come up with really new products and services? As an example the case of Red Bull® was analyzed and discussed. The question that immediately comes to mind when discussing innovation is simply which role R&D plays in this context. It was suggested that often successful and especially rapidly executed innovations are those that are simple and do not need an awful amount of R&D input.

- Is revolution possible in the food industry and who drives such revolution or at least evolution? The simple answer is: everyone in the company and innovation work is based on consumer insight driven by the business or technical capabilities coming from the R&D group or ideally from both partners.
- Another basic question was raised: is the food industry really innovation driven? It was suggested that the industry loves what is commonly described as *renovation* but is averse to real innovation. A historic list of food-related invention that changed the industry was discussed. The list shows the most recent inventions dating back to the 1950s! It is suggested that innovation in the food industry is almost "antique," and it depends on what is considered an important invention, such as hypoallergenic infant formula, which came into the marketplace in the early to mid-1990s.
- The need for real change in the industry was discussed and the two well-known examples of rather successful change (IBM) and not-so successful change (Kodak) were analyzed and discussed. Change or perish! And such, change always has an important influence on the structure and work of the R&D group.
- A rather important change to the basic business model of the food industry was suggested, namely a transition from B2C (business to consumers) to B2B (business to business, to the trade, the distributors). This transition would simply acknowledge today's reality and would put the dearly needed insight in the hands of the R&D group, namely customer insight.
- What's R&D worth? This is not a simple question and the perceived and real value of R&D are most of the time not the same. Two lists of arguments why R&D is either useless or useful were proposed and discussed. Consensus on these arguments is difficult to achieve and largely depends on your position in the company, both background as well as hierarchy. Ideally R&D should cost little to nothing and should bring big to super big to the company. It was suggested that one of the important reasons why food companies pay for an internal R&D organization is tax related. There are of course many other good reasons, not least of all the need for controlling critical core expertise inside the company.
- One of the most criticized downfalls of most of the R&D organizations in the industry is related to the fact that many or most members of the R&D group lack a sense of urgency and are too slow in their responses to the business. This is a truism that I can confirm, and it's most frustrating to see that there is not much progress to be observed.
- The chapter was terminated with a short discussion on why many companies always feel the urge or need to cut down on budgets, continuously "manage cost" (i.e., reduce them) and are often without good answers when the going gets tough. It is suggested that because the industry at large never had big crisis moments (such as the automotive industry in the United States only a few years ago, or the case of Kodak), its managers were mostly good-weather managers. When real difficulties show up, good-weather mangers try to do the job of bad-weather managers and they are mostly not up to the job. Moreover, self-critique of bad situations should not be silenced but needs to be heard and bad situations need to be addressed smartly and not in panic fashion; anticipation is the name of the game!

REFERENCES

"Food and food-related inventions," retrieved from http://www.enchantedlearning.com/inventors/food.shtml.

"The Last Kodak Moment?" 2012 January 14, *The Economist*, retrieved from http://www.economist.com/node/21542796.

Traitler, H, Coleman, B, Hofmann, K 2014, *Food industry design, technology and innovation*, Wiley-Blackwell, Hoboken, NJ.

Part 3

Disruptive outlook for the food industry's R&D

9 Outlook to other industries' R&D organizations

If we knew what it was we were doing, it would not be called research, would it?

Albert Einstein

9.1 INTRODUCTION

Let's take a look at how other companies and industries tackle R&D. I, Birgit Coleman, live in the San Francisco Bay area, and we have plenty of corporations across industries here locally and along the U.S. West Coast to learn from and to take away best practices. I will also venture outside of the United States for the sake of exercise to give us a holistic overview of what is going on in the secretive and open R&D worlds.

One thing is for sure: Food industry or not, R&D and innovation will be led by amazing talents, entrepreneurs, crowds, and crowd-sourced in community by the interaction between various disciplines and industries and by the unusual route and methods corporations will choose to take. I will share various angles of approaches so we can understand

- "How corporations can keep up with the speed of innovation?"
- "How corporations can join forces with startups?"
- "How innovation labs and teams can and should fit within the larger corporate structure?"
- "How corporates can reach and communicate with the consumer through new channels, technologies and just novel ways?"

There are some great corporations out there that have a 360-degree approach to R&D. We will take a look at these and then we will venture out into other categories and avenues (i.e., methodologies and processes) of success.

Food Industry R&D: A New Approach, First Edition. Helmut Traitler, Birgit Coleman and Adam Burbidge.

Besides all these various groundbreaking approaches, there is one underlying phenomenon and one common denominator: The future of any industry or any company is in the hands of computer geeks or simply technology. Why? No company can afford anymore to "just" offer products. The user, the end consumer, wants to have a personalized product, a service, or an experience with it. Such a transformation from a stand-alone product to a service offering happens through technology, which ultimately provides a service, a platform, an experience.

Let the inspirational journey begin.

9.2 BRIEF HISTORICAL REVIEW

If you think historically, go back 30 years or so, the organizations with big R&D divisions were AT&T, IBM, and Xerox; note that each of those companies had a de facto monopoly.

There are companies that invent revolutionary products, but someone else may commercialize them. Take for instance the integrated circuit that was independently developed by two companies, Fairchild Semiconductor and Texas Instruments, but Intel took the market from both of them. Whereas the graphical computer interface and mouse were invented and refined by companies like Xerox, then popularized with Apple's first Macintosh computer but Microsoft won the personal computer market space. There is no easy answer to the Xerox example of coming up with a great idea that someone else turns into a successful product. Companies have tried to steer away from the sort of fundamental research for which people win Nobel Prizes but instead take a more hybrid and applied approach and fast-forward to X approach.

That does not mean that we should abandon research and invention because of the challenges of commercialization. It is like saying, we should abandon agriculture because we do not know how to make delicious meals with the produce. The right answer is not to stop farming but rather to learn how to be a good cook.

This is exactly what this chapter is focusing on. Today we are at a point where we can draw ideas and solutions from a huge toolbox and technologies, which some short 30 or so years ago were not available, and we have Moore's law on our side, which predicts an exponential increase in computing power. The simplified version of this law states that processor speeds, or overall processing power, for computers will double every two years.

9.3 LET THE JOURNEY BEGIN: WHAT WE CAN LEARN FROM THEIR PLAYERS AND INDUSTRIES

9.3.1 Google

Let us start the journey with one of the continuously most innovative companies.

Fast Company, a business magazine that focuses on technology, business, and design, comes out every year with "The World's 50 Most Innovative Companies"; Google was the winner in 2014, in 2015 Google ranked #4 behind Warby Parker, Apple, and Alibaba.

You probably have heard of the new structure? G for Google; Alphabet Inc. will replace Google Inc. as the publicly traded entity. What is Alphabet? In Larry Page's letter (2015), he explains it the following:

> *Alphabet is mostly a collection of companies. The largest of which, of course, is Google. This newer Google is a bit slimmed down, with the companies that are pretty far afield of our main internet products contained in Alphabet instead. What do we mean by far afield? Good examples are our health efforts: Life Sciences (that works on the glucose-sensing contact lens), and Calico (focused on longevity). Fundamentally, we believe this allows us more management scale, as we can run things independently that aren't very related.*

The two co-founders Sergey Brin and Page are seriously in the business of starting new things. Alphabet will also include their X lab, which incubates new efforts like Wing, their drone-delivery effort. They are also stoked about growing their investment arms, Ventures and Capital, as part of this new structure.

9.3.2 Google X

The distinction between Google Research and Google X or X, which is what those who work there usually call it, is sometimes framed this way: X is tasked with making actual objects that interact with the physical world that needs to meet three criteria.

1. All ideas and projects must address a problem that affects millions or even better, billions, of people.
2. All initiatives must use a radical solution that carries at least one science fiction component.
3. All must tap into technologies that are now or nearly obtainable.

Last but not least and needless to say, no idea should be incremental.

One of the explanations for what the X actually stands for is the search for solutions that are better by a factor of 10 or to build technologies that are 10 years away from making a large impact. Coming back to incremental ideas; it is hard to do or launch almost anything in this world. So, why not shooting for the moon and revolutionizing the world? Attacking a problem that is twice (incremental) as big or 10 times as big is not twice or 10 times as hard. It is often the opposite. Making inroads on the biggest problems is often easier as you have the privilege to start over, to re-exam what your product or service actually is and solves, or to get to dump the idea of your original product or services altogether in favor of a substitute…and then you might come up with something worthy of X.

Side note: Singularity University

Yes, it reminds you probably of the works of the Singularity University (SU) and their activities.

Singularity University is a California Benefit Corporation part university, part think-tank, part business incubator located in Silicon Valley whose stated aim is to "educate, inspire and empower leaders to apply exponential technologies to address humanity's grand challenges." It was founded in 2008 by Peter Diamandis, Ray Kurzweil, and Salim Ismail in NASA Research Park, CA. Singularity University aims to develop a global network of innovation ecosystems alongside a coalition of entrepreneurs, business leaders, universities, government agencies, and nonprofits. The institution taps into a large and growing community of scientists, thinkers, engineers, investors, business leaders, and public policy makers who are motivated to explore the potential of rapidly advancing technologies to take on humanity's big challenges, such as water scarcity and energy consumption.

Among their services are their annual conferences focused on exponentially accelerating technologies that are impacting individual verticals such as Finance, Medicine & Healthcare, and Manufacturing, and the Singularity Hub, an independent science and tech media Web site published by SU. Singularity University also opened an accelerator called SU Labs to accommodate start-ups which aims to "change the lives of a billion people." Up to now, it has also coached roughly 30 startups such as Made in Space, which has developed a three-dimensional printer adapted to the constraints of space travel. The first prototype of Made in Space's the Zero-G Printer was developed with NASA and sent into space in September 2014. In 2011, a Singularity group launched Matternet, a startup that aims to harness drone technology to haul goods in developing countries that lack highway infrastructure. Among several startups springing from Singularity are the peer-to-peer car-sharing service Getaround and BioMine, which uses mining industry technologies to extract value from electronic waste. You clearly get the similarities in the mind-set to harness the power of innovation.

Here is the fundamental challenge of spinning out extreme solutions to big problems for any company: society tends to move incrementally, even as many fields of technology seem to advance exponentially.

9.3.3 Back to Google X and the future

Probably where Google X's unique approach lies is that they are ready to take incredible risks (and massive amounts of funds) across a wide variety of technological domains and to not hesitate to stray far from its parent company's business with their unimaginable army of talents. Secondly, Google is hitting perfect timing and is situated in a perfect intersection where networks meet computing power and artificial intelligence.

Let's look deeper inside the X and its projects.

First of all, it is exceedingly rare that a new Google X project is created. It is almost an art to define a problem. Some problems are easier to see in the rearview mirror—try explaining to your pre-smartphone self how much x or y is going to change your life—and for some challenges it is a matter of looking back from the future and it becomes obvious that I want for example to be connected to information in a minimally invasive way.

Work at the lab is overseen by Sergey, while scientist and entrepreneur Astro Teller (Captain of Moonshots) directs day-to-day activities. The lab started up in 2010 with the development of a self-driving car. Google X projects are often referred to as "moonshoots" within the company. For example, Calico, Google's life-extension project, is considered a moonshoot but is not part of Google X. The same is true of Google's project to build robots for businesses. You probably have heard already of projects such as "Project Wing" that aims to rapidly deliver products across a city using flying vehicles, similar to the Amazon Prime Air concept. "Project Glass" is another one that develops an augmented reality head-mounted display, or "Project Loon," a high altitude Wi-Fi balloon that aims to bring Internet access to everyone by creating an internet network of balloons flying through the stratosphere. Mostly, X seeks out people who want to build actually physical things.

Feeling proud and accomplished having identified and defined a problem?

Meet Google X's Rapid Evaluation Team or "Rapid Eval," where ideas get vetted and tested out primarily by doing everything humanly and technologically possible to make them fall apart. For the exact following and perfectly framed reason by Rich DeVaul, Head of Rapid Eval (Gertner, 2014):

"Why put off failing until tomorrow or next week if you can fail now?"

Failure is not the goal at Google X but in many respects it is the means.

9.3.4 Google Research

Let's meet Google's "sister": Google Research.

It is devoted mainly to computer science and Internet technologies. Google claims its approach as a hybrid model: to improve Google products and services and through the broader advancement of scientific knowledge. In this model and in practice, it blurs the line between research and engineering activities and encourages teams to pursue the right balance of each. It is not only the right balance between research and engineering, but it is also a healthy mix of short- and long-term research projects. Examples of multiyear, large and complex systems efforts include Google Translate, Google Health, Chrome, among others.

In the paper "Google's Hybrid Approach to Research," Spector and colleagues (2012) explain Google's way of research as:

First of all, it happens throughout Google, exploring technical innovations whose implementation is risky. Sometimes, research at Google operates in entirely new spaces, but most frequently, the goals are major advances in areas where the bar is already high, but there is still potential for new methods. Its methodology is iterative and usually involves writing code from day one.

In the same paper and based on their volume of R&D experience and experiments, Google documented few research patterns to better navigate the minefield of resources needed after research projects are launched and depending from what starting point:

- Engineering → Research: Transitioning a project from an engineering team to the research team is an important mechanism for giving a project more time or resources.
- Collaborative Integration of research and development teams → Engineering to follow-up.

- Traditional Research and Development model: A project in the research group that creates new concepts and technologies, which are then applied to existing products or services.
- Research → Operation: A project in the research group that results in new products or services based on that research.

9.3.5 Google for Entrepreneurs (GfE)

There are campuses in cities like Warsaw, São Paulo, Tel Aviv, Madrid, London, Seoul, and, where you can dive into your local community and where you come together to learn, share ideas, and launch your startups within their sets of tools and platform within an amazing ecosystem. Have you heard of these campuses and outposts?! Cannot blame you? It's basically a soft-branded Google effort to support entrepreneurship.

Although the GfE is not to be mixed up with the Google Developers' Launchpad that provides you with the technology, events, online resources, expertise, and community to launch and scale your app. The Launchpad program leverages Google's infrastructure, together with UX reviews, mentorship, and events, to let startups anywhere focus on building, distributing, and monetizing your app, everywhere. Successful applicants join their program in one of two stages: Scale or Start. The respective team reviews each application individually and in context of country, industry, technology, growth trajectory, and business model (e.g., consumer app or business-to-business). Startups in the Scale tier usually have established product-market fit, membership in an incubator or accelerator that Google works closely with, referrals from "Googlers" in the Developer Platforms group and favorable local media coverage or reputation. If you have validating your idea, and have tested a prototype, you take the next step to get user feedback and develop your product with Launchpad's resources.

9.3.6 Google Ventures

And then since 2009, there is Google Ventures (GV). It provides seed and growth-stage funding to bold new technology companies. They have backed more than 300 companies, including Uber, nest, slack, Flatiron Health, just to name a few. Their investment strategy is in every field, but they have a unique focus on machine learning and life science investing. To make things clear, GV does not make strategic investments for Google. But their extraordinary hands-on work with startups is worth looking into closer because they are known for their Service model. GV was one of the first venture capital (VC) firms to offer portfolio companies support in design, recruiting, marketing, and product management. Partners engage top designers and marketing partners from outside, successful serial entrepreneurs, and of course, top engineers within the firm among other talents.

Let me share a special-developed offering by Google to their portfolio companies. In 2010, Google Venture design partner Jake Knapp began running design sprints at Google. Basically, a design sprint is an intense five-day design process where companies can build and test nearly any idea in just 40 hours. First, Knapp worked with people from Chrome, Google Search, and

Google X. In 2012, he brought sprints into Google Ventures who perfected the process with hundreds of their portfolio companies and even came out with a practical guidebook, *Design Sprint* in late 2015.

We will have another opportunity to learn more about the so-called "sprints," which Zappos.com, the online shoe and clothing shop, has applied successfully.

Speaking of Google and design, there has been some interesting observations made. Historically, design, the process of a creation, has been a qualitative art. It seems that we are witnessing a similar transformation in design thinking whereby it is becoming a quantitative science like physics, chemistry, and biology. Design thinking is ready to become a foundational science for engineering and innovation.

In the last two decades, breakthroughs in information technology and genetics have turned biology from science to a more quantitative and predictive one. As a result, biology has become a foundational science for engineering. There is the world's first synthetic biology accelerator in San Francisco called IndieBio, which is short for Independent Biology. The goal is to build a future where biology is not only a field of study, but a technology that will help solve our world's most challenging problems, from feeding a growing population, to providing energy, to treating or curing the maladies that limit or kill us. IndieBio provides seed funding and intensive mentorship for a three-month transition into the world.

If you seek for inspiring new business or design-thinking processes or models for that matter, look no further than the retail and fashion industry, an industry that gets disrupted as we speak. Shopping is now a highly customized and social experience.

9.3.7 Westfield Labs: Designing the mall of the future

Staying with the theme of *Fast Company's* most innovative companies, Westfield labs scored a rank for the first time, #36, to give the mall a high-tech makeover. The Westfield Group owns and operates 40 shopping centers around the world. In 2012, the company established a testing lab inside the Westfield San Francisco Center to reimagine the retail ecosystem and to drive the future of shopping and retail. After years of show-rooming and online retail with immersive personalized mobile experiences, the data-driven shopping experience lands inside brick-and-mortar stores. The lines between the physical and digital worlds are blurring.

One big challenge for the Lab is to understand how digital and brick-and-mortar shopping experience can go hand in hand and to find solutions by running various high-tech experiments. Although in-store shopping still accounts for the bulk of revenue for retailers, e-commerce is catching up. Westfield Group wants to give the traditional commerce companies a safe environment to test different types of technology. Since inception, the group launched various experiments. One is to offer brands to lease small spaces inside the mall to test state-of-the-art technologies and new design elements. Another one, called Bespoke, offers a place that functions as a co-working, event (e.g., for fashion shows and hackathons) and technology demo space all under one roof.

But perhaps the most press got Westfield Labs was through a partnership with eBay. With customers being able to buy online or offline, globally and locally, eBay Inc. is particularly interested in its latest shopping innovation. A trio of large, interactive digital storefronts in the Westfield shopping mall in San Francisco gave consumers the opportunity to swipe through and buy curated merchandise from each brand—Sony, Rebecca Minkoff, and TOMS Shoes—via a "shoppable window," pay with PayPal, and then arrange for free home delivery or pick up.

Sounds fabulous, right? What the two partners learned though from this pilot was not to make these screens "shoppable" but rather to keep them as displays only showing images of products from nearby stores and where to find them. Why? People visiting the mall want to actually buy something and get the instant gratification. Still, these large digital displays generate more sales from existing space and the two partners were able to create a hybrid physical commerce sector and not just online or offline transactions.

With Bespoke and Westfield Labs, the company is encouraging innovation in online and offline shopping. By letting brands lease small spaces inside the mall to test state-of-the-art technologies and new design elements, they are giving traditional retailers a safe environment to experiment with the technology of the future.

9.3.8 Attack on the the brick-and-mortar model by e-tailers Zappos and Amazon

Zappos.com is an online shoe and clothing shop based in Las Vegas, Nevada. It was founded in 1999 when founder Nick Swinmurn could not find a pair of brown Airwalks at his local mall. Fast-forward 10 years and Amazon announced that it was buying Zappos for $940 million in a stock and cash deal. But the amazing part of the deal was that Amazon was and still is open to let Zappos continue to operate as an independent entity keeping its headquarters in Las Vegas. In 2010, Zappos opened an office in San Francisco called Zappos Labs with their tag line "Exploring the Future of Zappos," which could include crazy retail-related experiments or new potential lines of business.

Known for its customer service, the stories of just how far Zappos will go to make a customer happy are legendary. Zappos's customer service has included quick and easy refunds, sending customers flowers to cheer them up and speaking with a customer for more than 10 hours, just because that person needed someone to talk to. This is pretty much unheard of. So, it also did not come to as a surprise when Zappos introduced a new service in 2014 called "Ask Zappos," a digital personal assistant who takes requests in the form of images and finds the exact item while also providing links to some alternatives (see image 9.1).

It reminded me somehow of *Foodspotting*, which is a community-driven dish discovery app that allows individuals to find, share, and recommend food dishes, or the music-identifying Shazam Entertainment app, which is an amazing way to discover, explore, buy, and share music. Say you are in a restaurant or a store and a song comes on and you cannot quite place the tune. In the past, your options were limited, you could try asking somebody in that store or restaurant for a clue or hum the song in front of your partner or friends and make a fool of yourself. But with Shazam you press a button on your phone and in seconds you will get the artist and the song title.

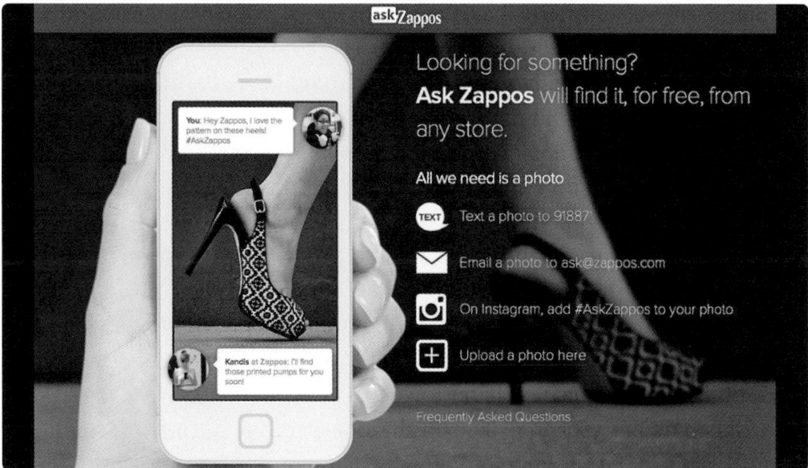

Image 9.1 Ask Zappos. Source: Blogpost "The Customized, Social Future of Retail," Birgit Coleman, San Francisco, September 7, 2015.

9.3.9 The rise of social shopping

Speaking of Amazon and of the future of shopping, social shopping is picking up steam. Social shopping is a way of shopping that includes the shopper's friends and technology is the medium that mimics the physical mall and stores.

In mid-2014, Amazon and Twitter introduced a new feature that lets users add items to their Amazon carts by including a hash-tag within a tweet. "Add it now, buy it later" is the slogan Amazon is building on. For Amazon, extending its reach into social hubs is an obvious strategy to defend itself against the rise of alternative social retail hubs such as Pinterest or Instagram. Another great side-effect is that Amazon gives product makers a nice "pull" to sell their stuff on Amazon to tweet out product links that are identifiable as being sold on Amazon, which is a prerequisite for the #AmazonCart feature to function. This is also not to mention the free advertising powered by Twitter users.

A few words on Pinterest and Instagram emphasize the power of social. Pinterest is probably the biggest site where consumers share their wants, desires, and intentions. Instagram, meanwhile, gives users the ability to follow and interact with the shopping exploits of their friends, family members, and role models.

In 2015, both rolled out new features: Pinterest's Buy it button and Instagram's Shop Now. And once Twitter more widely expands its Buy button, it will significantly shift that social network's role in the relationship between consumers and businesses and further disrupt the way consumers think about path-to-purchase.

Companies like Free People, an American clothing retailer, are tapping into the power of social engagement for their brand and stores. A few years ago, Free People invited fans to upload pictures of their denim purchases to Instagram using the hashtag #MYFPDENIM. This kind of crowd-sourced campaign lets them aggregate photos onto a microsite. For consumers, they get to be a part of a campaign in which the best photo of the day is featured prominently online.

Zappos Labs has conducted online shopping experiments, including a curated digital magazine, a recommendation engine based on Pinterest pins, and personal style recommendations based on Instagram posts. Dubbed *Pinpointing*, a Pinterest Shopping Companion Tool, Zappos makes product recommendations based on what a user has pinned and products that had been pinned recently.

Many new social shopping marketplaces such as Shopa, Shopcade, The Hunt, and Net-a-Porter's Net Set have emerged as smartphone and tablet adoption have risen. London- and New York-based Shopa, launched in 2012, rewards users with discounts for promoting products they like. Peter Janes, Shopa's Founder and CEO, says,

> *The technology allows us to track everything that happens with regards to our retail partners products across any platform and device. Then we can feed that back to retailers. Basically in real-time we can show them who the top advocates are, which products are trending in real-time, which platforms deliver the best ROI, and we can track the whole process from a share to a purchase (Butcher 2015).*

According to Shopa, some studies suggest that a social recommendation of a product is 45 times more effective in generating sales than any traditional marketing or advertising method.

With the popularity of Net Set and cult classics like The Hunt—a community-driven shopping experience that tracks down the items you covet but don't know where to find—retailers are well on their way to encouraging more spending and to collect even more data on its burgeoning millennial customer base thanks to social e-commerce.

Net Set users can shop within the app, comment on styles they like, recommend fashions to friends, take a voyeuristic virtual peek into the closets of stylish women, follow a Style Council of 15 fashionistas, and join or create Style Tribes, which are groups within the app with common interests or aesthetics. The app also boasts image recognition, the holy grail for e-commerce.

So, just when we feared the death of brick-and-mortar stores, when we thought shopping malls were becoming passé, and online shopping was getting dull, retailers are transforming. I would even daresay that as a change agent I would never ignore observing the fashion world as you get a good reading on what is happening out there and as we will hear more so in the next two chapters, the retailer.

What we should take away from all these examples of experiments is that it is that beautiful dance or in business terms "innovation partnership" between corporations, startups, academia, end-consumer, retailer, and new and established technologies and business partners that makes us succeed.

Let's go back to the new and previously mentioned feature "Ask Zappos". The holy grail does not only lay in the collaboration with various young or established partners across industries but also internally across disciplines. This service was created start-to-finish in 12 weeks, using a process of "one-week sprints," which is part of an Agile methodology of project management called *Scrum*.

This is how generally a sprint works: a team commits a set amount of work within the week. After an initial planning meeting, there are daily 5- to 10-minute scrums or standing meetings where the team quickly discusses what they have been working on, what challenges they face and what the next steps are. Those meetings are tactical; the end-of-week retrospective meeting is a reflective one on the process and what worked and what didn't, which is as valuable as a process.

The beauty about this whole process is, that it is a pull-versus-push mechanism. Teams are committed and they hold each other accountable. The individual steps give room for flexibility in terms of changing requirements. Zappos conducts costumer interviews, and based on that feedback, they will adapt it in their sprints. It is not about being absolutely perfect for 100,000 users but to let 100 users try it out and then fix things.

9.3.10 Traditional industries meet tech

The art of shopping and the retail industry are not the only areas and industries that get disrupted that we can learn from. Any industry has been or is going through their next transformation as a result of technology is a good place to start. Again, the fashion industry is just a great example how it has to find ways to collaborate with new emerging sectors such as wearable technologies combined with new kids on the block such as high-fashion retailer Net-a-Porter. It is the art of successfully blending content and commerce, delivering it in a distinctive way with close attention to service and customer care. On this note, one more example in the social shopping fashion space because it is a phenomenal example, betabrand. Betabrand is a San Francisco-based crowdfunded clothing community. Basically, brand-new ideas spring to life every day—yes, every day. Their fans co-design and crowdfund them into existence in a matter of weeks. How does it work? Voters decide what betabrand should make next. Voters get first dibs at the biggest savings when a concept goes in crowdfunding, and they get early access by wearing it first when they fund a prototype.

9.3.11 The art of dating

Let's venture into another traditional industry and see what we can learn from: Dating. Dating is defined as a part of human mating process whereby two people meet socially for companionship, beyond the level of friendship, or with the aim of an intimate relationship or marriage. Fast-forward to 2015, and we have a $2 billion online dating industry.

The market leader in Web dating is InterActiveCorp (IAC), which owns at least 30 sites, including Match.com, OKCupid, and eHarmony.com. After OKCupid was bought by IAC for $50 million in February 2011, it decided to set up an innovation hub in San Francisco to hash out the site's next steps but also to make corporate entrepreneurship work within IAC. The goal of OKCupid Labs is to launch lots of projects and to give Labs employees a safety net to experiment, which comes with the Silicon Valley territory, "Fail Fast, Fail Often." This open mind-set should pave the way to venture into maybe entirely new dating businesses or ways to bring people together that is not explicitly dating. What will happen if we take the word *dating* out of the equation and re-name it *XYZ*? What new business venture would emerge?

9.3.12 Learning from the least sexy industry role model

Never underestimate one of the least sexy industry or space to innovate and its potential of opportunities. Intuit Inc. is a software company that develops financial and tax preparation software and related services for small businesses, accountants, and individuals. It is headquartered in Mountain View, CA. Intuit Labs is Intuit's literal Innovation System or Machine. The amount of resources and tools the company puts on the table for their employees and the array of projects they are spinning off and have tested are mind blowing.

I personally love their "Incubation Week" set-up which is a five-day immersive experience where Intuit teams go from idea to MVP with resources starting from expert coaches and executives across the firm to top design talent, expert engineers, legal and compliance consultants to free food and fun to turn your side project into reality. One time the Incubation Week was applied to better understand their customers and to transform the way the outbound calling/ sales powerhouse does business. The product and the market had changed but they hadn't adapted the business model.

If it wasn't for Incubation Week, a lot of projects would not have been born. Projects such as QuickBooks for Bitcoin Payments or one dubbed *swyft* thinking of a hands-free payment experience for merchants by leveraging wearable devices such as Google Glass or a free app called *Keep* that lets you put money away that you normally waste toward the things you care thanks to helpful reminders and visual rewards every time you save money. Also, at the end of each incubation week, Intuit hosts a Gallery Walk, which serves as an opportunity for the teams to present what they have accomplished. This is a great way for other employees to better understand what to expect from such an Incubation Week and to get motivated to form or to join a team and also to see what the company is working on and to share that internal knowledge.

There is much more to be learned from Inuit Labs and their internal innovation catalysts or their Demandforce salespeople. Intuit was a partner also at our TEDxSanFrancisco event in October 2015. I intentionally didn't call Intuit or any company as our sponsor because our goal with the TEDx event was to turn sponsors into partners and let the audience experience the brand rather than create a brand cemetery through tons of signage throughout the event location and day. The classic sponsorship model is passé. We want to get to know the brand; we want to learn from it, play, experiment with, and experience it. I am even going a step further to state that the new consumer not just wants a personalized brand experience and experiment with it but also to see it socially responsible or active. I mentioned previously TOMS Shoes, which I would like to quickly mention in this category. TOMS's "One for One. Improving Lives" slogan and initiative is making a difference. With every product purchased, TOMS Giving Team provides shoes, sight, water, and safer birth services to people in need together with Giving Partners.

Indiegogo would have been another poster child for transforming ideas worth spreading, which was the TED mantra, into actions and causes. Indiegogo is an international crowdfunding platform founded in 2008 and headquartered in San Francisco. It allows people to solicit funds for an idea, charity, or startup business to finance their goal. It addresses the need and audience of the classical "technology adoption life cycle model" as described by Ryan and Gross: Innovators - Early Adopters - Early Majority - Late Majority - Laggards.

The technology adoption lifecycle model describes the adoption or acceptance of a new product or innovation, according to the demographic and psychological characteristics of defined adopter groups. The process of adoption over time is typically illustrated as a classical normal distribution or "bell curve." The model indicates that the first group of people to use a new product are *innovators*, followed by *early adopters*. Next come the early and late *majority*, and the last group to eventually adopt a product are the *laggards*.

For any startup, receiving vital feedback on your first product or product line, gaining your first customers, and creating brand awareness, is critical. The rest will just follow automatically, once you accomplish having the early adopters on board or building buzz around your brand to secure, for example, your first major distribution deal.

The plan was, together with Indiegogo, to invite and encourage the TEDxSanFrancisco audience to collectively identify one of our speakers' projects or causes and to announce and devote a crowdfunding campaign after the talk directly from the TEDx stage. To help you understand how we nominate, let me share the key factors to this decision:

- Social relevance in local community
- Global social relevance
- Can it be replicated to other communities?
- Does it solve a global challenge (nutrition, environment, water, space, education, extreme poverty, infrastructure, etc.)?

Our TEDx social channels will keep these campaigns up and running even after the conference day; I cannot wait to see the event unfold and to unleash its potential by giving the audience a trusted platform. Unfortunately, we were at crunch mode for the TEDx preparations at this point of writing this chapter, but I hope we will leave a community, a legacy and tangible projects, and outputs.

9.4 HALFTIME

So far, we have looked into disruptive and traditional industries and the world's most innovative companies, mostly native Silicon Valley corporations, that serve us as innovation role models. Many international corporations have taken the active step to set up shop—be it in the form of an innovation lab or technology outpost or digital media hub—here in the San Francisco Bay Area across all industries, starting from Nestlé or Swisscom to Citrix Labs, American Express, Honda, Siemens, and actually you name it. The next part of the chapter, I would like to focus on other avenues of R&D approaches.

9.4.1 The lean startup methodology

The Lean Startup methodology is a method for developing businesses and products first proposed in 2011 by Eric Ries. He claims that startups can shorten their product cycles by adopting a combination of business-hypothesis-driven experimentation, iterative product releases, and what he calls "validated learning." The lean startup philosophy has a premise that every startup is a grand experiment that attempts to answer a question. The question is not "Can this product be built?" Instead, the questions are "*Should* this product be built?" and "Can we build a sustainable business around this set of products and services?"

This experiment is more than just theoretical inquiry; it is a first product. By the time that product is ready to be distributed, it will already have established customers. It will have solved real problems and offers detailed specifications for what needs to be built.

The first step is figuring out the problem that needs to be solved and then to develop a minimum viable product (MVP) to begin the process of learning as quickly as possible. Once the MVP is established, a startup can work on tuning the engine.

The Lean Startup philosophy became widely popular with Silicon Valley's tech startups and several prominent high-tech corporations such as Intuit, Dropbox, GE, and Wealthfront have also begun to live out that philosophy.

9.4.2 The lean network approach: The nomad approach

I see a new trend with the innovation labs or technology outposts that international companies setup early on. I like that companies question their current setups and ask themselves if their current setup is sustainable and the most efficient business model to generate R&D, innovation, or launch pilot projects. I see more and more corporations using the lean startup model, a lean-network approach.

If you apply this lean approach to your external network via satellite offices by downsizing on various outposts, instead have your employees scout for trends and technologies via a lean nomad lifestyle approach. Basically, corporations send one or the other employee to a location and the appropriate person will move through cities or countries depending on need and fit and will create the required community accordingly on site to cocreate innovation in a lean startup approach.

9.4.3 R&D-I-Y

There is one more R&D term we can add: R&D-I-Y, R&D-It-Yourself.

In Britta Riley's TEDxManhattan talk "A garden in my apartment" in 2011, she introduced the R&D-I-Y Web platform consisting of 18,000 international citizens collaboratively innovating various open source projects such as flagship project Windowfarms, a new modern kitchen garden. Windowfarms is a system first set up in any window that gets a few hours of light per day. Then, consumers select a plan choosing a flavor mix and size and the first crop gets shipped. Every month local farmers from Windowfarms systems will replenish the kitchen garden with old favorites, can send pro tips to build cooking repertoire, and can help add unexpected diversity.

But the path to get to the Windowfarms system right was not without many collaborative interactions and iterations. One volunteer windowfarmer suggested to replace the water pumps with air pumps to cut the carbon footprint in half. Another tester tackled growing experience pains when he brought strawberries to fruit for 9 months out of a year by chancing out the organic nutrients, and window farmers in Finland have been customizing their farms for their dark days of the Finnish winters by outfitting them with LED grow lights.

What a lean approach to hundreds of solutions. With this open source and crowd-sourcing approach, volunteers basically did the global and the local market roll-out without customer frustrations. Windowfarms' culture embraced testers who supported somebody else's idea and the community experienced the joys of collaboration across the globe. That is generally unheard of, to achieve a high customer satisfaction rate with a trusted and committed army of enthusiastic volunteers.

We will discuss in more detail the power and philosophies behind open source. Crowd-sourcing, the process of obtaining needed services, ideas, or content by soliciting contributions from a large group of people or online community, rather than from traditional employees or suppliers, is another phenomenon corporations should tap into.

9.4.4 The IKEA effect

The R&D-I-Y approach reminds me yet of another phenomenon. It proves that users have fun tinkering with their systems to adopt them to their individual needs. It's like the "IKEA Effect," which is the increase in valuation of self-made products. Studies show that customer satisfaction increases with their products after having assembled it themselves and they saw their amateurish creations as similar in value to the creations of experts.

That added value customers attach to their own labor is actually part of a broader trend underlying consumer involvement as companies have shifted in recent years from viewing customers as recipients of value to viewing them as co-creators of value. Companies now actively include consumers in the design, marketing, and testing of products; think in terms of hackathons, social media, and collaborative and community-built examples we have discussed for the ultimate and priceless goal: customer satisfaction!

Let's defeat our best enemy: the "not invented here" syndrome where managers refuse to use perfectly good ideas developed elsewhere in favor of their internally developed ideas.

9.4.5 Open source

We cannot neglect IP with this topic. Let us look at a few more corporations and how they approach IP in novel ways with or without a hidden agenda.

Open source means that an item is produced and developed openly and with no license restricting its design, use, or distribution.

In mid-2014, Elon Musk, CEO and CTO of SpaceX, CEO and chief product architect of Tesla Motors, and chairman of SolarCity, announced the releases of Tesla's patents to "Good Faith" Use with the statement "We believe that Tesla, other companies making electric cars, and the world would all benefit from a common, rapidly-evolving technology platform." Let's also call it a win–win but also an upside for Tesla. It possibly would aide Tesla's rate of adoption because it may encourage other companies to build charging stations and other products that would support Tesla's growth. Just to make sure also, that anyone who thinks that Tesla just abandons all its IP, the company still owns a wealth of IP in the form of trademarks and trade secrets. Anybody that decides to copy Tesla's cars too closely, may in fact be sued by Tesla.

Tesla is by far not the only and first contributor. Although giants like Google, Twitter, IBM, and Netflix contribute a lot of open-source software, Facebook has taken open source to its logical extreme. The company open sources everything: software, hardware, and know-how. In 2014 alone, Facebook launched 107 open-source projects and actively contributes to many external projects, including Apache Hadoop and MySQL. On the hardware side, dubbed the Open Compute project, Facebook revealed all details necessary to companies building their own Facebook-like data centers.

In part, this is an excellent way to attract developers to its platform, and that leads us to another bottom line. Let's not forget the big picture: Technology leadership is not defined by the number and content of patents but rather by the ability of a company to attract and motivate the best talents. We live in an age of mass competition where the ability to land top talent is the success of the company or the product itself. Success lies in the ability to innovate, to execute, and to engage with a vibrant community.

But we might have a new kid on the block joining the electric car community with an (closed source) Apple-branded electric vehicle, codename "Titan."

9.4.6 The street is your R&D lab

When resources like capital, natural resources, and services are scarce, you need to be creative or in other terms, you have to be *jugaad*. Jugaad is a colloquial Hindi-Urdu word that can mean an innovative fix or a simple work-around. It is time to improvise and create solutions and to become an alchemist. It is the art of doing more with less.

In some cases and countries, jugaad solutions are the secret weapons of emerging markets and are even leap-frogging the nod. China, for example is using telemedicine instead of building luxury hospitals. In Silicon Valley or generally in the West, we always chase the next thing. But more for more is not sustainable. We cannot longer afford this approach. Often, in Corporate America, we call jugaad *frugal innovation* or *reverse innovation* or *BOP (Bottom of the Pyramid)* innovation. Western CEOs and companies need to perceive scarcity as an opportunity to innovate and leverage employee's ingenuity to create faster, better and cheaper solutions. The core principle Jugaad innovators operate on is that they leverage partner networks.

For example, you don't design solutions by sitting in an R&D or innovation lab. Instead, collaborate with local partners to learn about local problems and identify end user's needs. Take portable infant warmer for premature babies, Embrace, co-founded by Stanford graduates. The device does not need electricity and helps mothers hold their babies close to their bodies, boosting their survival. Embrace used NGOs and hospitals in emerging markets to understand end user's constraints like lack of electricity or GE Healthcare to scale up the distribution of its infant warmers.

Secondly, do not reinvent the wheel but leverage existing resources. For example, co-develop frugal and sustainable solutions by engaging local communities or global partners to design and even build your next offering. MittiCool, a fridge made entirely out of clay that consumes no electricity, draws on this principle.

9.4.7 Projects to promote interdependence

Another movement and similar to jugaad's philosophy, to discuss is the phenomenon of a shared economy or also referred to as a collaborative or mesh economy. It is a socioeconomic system built around the sharing of human and physical resources. Participants share access to products or services and a community (people) rather than having individual ownership and resulting in new communities and opportunities. That's how entire new product groups, business models, and companies emerged, such as social lending, peer-to-peer accommodation and travel experiences, task assignments, and transportation sharing.

Corporations have a couple of choices: It is either to innovate with that co-creation and sustainable mind-set, which often requires cannibalizing your own business or business model, or to be eaten alive and disrupted, which could result in losing market share and revenue, including losing your innovative edge that could ultimately cost you your talent, which is among your biggest and most precious asset.

9.5 SUMMARY AND MAJOR LEARNING

We have heard and learned from many corporations how they tackle R&D in many novel ways. More and more companies operating in many different fields of industries establish separate innovation and technology outposts in the San Francisco Bay Area to be close to the entrepreneurial activities, talents, and mind-set. Firms provide an array of resources and tools for their employees to create and engage in a risk taking and open innovation culture and to attract the best talent. But what makes one company succeed and the other less or not?

A 360-degree innovation approach is on the one hand, the art of incorporating and balancing a healthy mix of these discussed methodologies and tangible and intangible setups to bring innovation to the customer. It is on the other hand, to also engage with your outside world, your local community, and your local/regional startup. It is great if you can attract and retain your best and most motivated talent but you cannot do it all on your own and just in house.

REFERENCES

Butcher, M (2015), "Social Shopping App Shopa Secures $11M Series A to Scale Internationally," *Tech Crunch*, retrieved from http://techcrunch.com/2015/02/17/social-shopping-app-shopa-secures-11m-series-a-to-scale-internationally/

Gertner, J. 2014, April 15, "The Truth about Google X: An Exclusive Look Behind the Secretive Labs Closed Doors," *Fast Company*, retrieved from http://www.fastcompany.com/3028156/united-states-of-innovation/the-google-x-factor.

Musk, E., 2014, June 12, "All Our Patent Are Belong to You," retrieved from https://www.teslamotors.com/blog/all-our-patent-are-belong-you

Page, L., 2015 "G is for Google," retrieved from https://abc.xyz/investor/founders-letters/2015/index.html#2015-larry-alphabet-letter.

Ries, E., 2011, September, "The Lean Startup: How Today's Entrepreneurs Use Continuous Innovation to Create Radically Successful Businesses," *Crown Business*.

Spector, A., Norvig, P., and Petrov, S., 2012, "Google's Hybrid Approach to Research," *Communications of the ACM*, doi: 10.1145/2209249.2209262

10 Utopia or visions for the future: A new reality?

It's illegitimate to talk about a post-scarcity Utopia without talking about questions of distribution. There have always been these Utopian predictions—"electricity too cheap to meter" was the atomic promise of the 1950s.

Mitch Kapor

I may not have gone where I intended to go, but I think I have ended up where I intended to be.

Douglas Adams

10.1 WHAT IF I HAD A MAGIC WAND? MY FIRST SET OF MAGIC TRICKS

Ever since Harry Potter and his magical capabilities became so popular I was intrigued by the following questions: What if I had a magic wand? What would I do with it and what tricks or actions would I perform? I am not talking about all these stupid ideas that I would not want to repeat here but what great stuff—apart from world peace—would I go for and make it happen? I mean, great stuff in the professional and business area, although a bit more craziness would do the business side an awful lot of good. I will also perform a few crazy magic tricks, or at least I will try. Some of you may know the magician duo Penn and Teller and the really entertaining show on TV *Penn and Teller: Fool Us*. Magicians, young, old, new, and inexperienced and more experienced ones do their stunts and Penn and Teller have to find them out; most of the time they do, showing their great width and talent in this business, and it's funny and entertaining at the same time, which of course is the purpose of the show. Let us do something similar here: let's play *Penn and Teller: Fool Us*. I admit that it's a bit tricky, as for the purpose of this chapter, I will have to be the one who wants to fool and at the same time I have to be Penn and Teller and check out my own magic trick. For instance, there's the one that makes a food

Food Industry R&D: A New Approach, First Edition. Helmut Traitler, Birgit Coleman and Adam Burbidge.

company, especially a large one, an entirely R&D-centric and R&D-driven enterprise being so much more successful than a comparable, traditionally organized company today. I shall discuss this in more detail, but let me first list a few more tricks I would like to perform.

10.1.1 Abracadabra...

And then there's this other magic trick that would transform the present business model of the food industry in which the "customer is king," and the customer is the trade, the distributors, the supermarkets, or simply every organization that sells food and beverage products to consumers. In the not so distant past, supermarkets, the typical large scale outlets for food and beverages, were seen by the manufacturing branded food industry in antagonistic and negative ways, almost like enemies. Most of us in the industry could observe a substantial improvement of this situation, however, there is still much room for improvement. So much so that I would magically change the situation to one that really puts the two sides in the same playing field, almost as if they were one company, or if retailers were an almost integrated extended arm of the manufacturing company. For some retailers, especially those who sell their own store brands, this already is reality, at least to some degree. Also in this case, there is room for betterment especially with regard to opening up to a much larger playing field, including many more manufacturers and distributors.

I am not advocating the "single provider" model for the food industry. I do realize that not only would it most likely not work, but antitrust laws and lawmakers would shatter such a model in no time. But let me just expand on the integration part of the industry, which ideally could and should be driven by the R&D branch, similar to what you already find in the automotive or computer industry. What I mean by that is the following example from the car industry: safety relevant items such as air bags or ABS brake systems had to be introduced across the entire industry within a short period of time after their inception. Other examples from the computer are microprocessors from one company in many or most of today's computers, be it Microsoft, Apple, or other operating systems; almost all of them have microprocessors from the major manufacturer (Intel). I could list more examples from many more industries, but my point here is simply integration of critical components or building blocks for different products, ultimately assembled by the "last in line," can enhance efficiency and speed of execution and adds to safety, quality, and cost efficiency for the final client or consumer.

10.1.2 Integration across the borders in the food industry

Now, how is this done in the food industry? There are some elements of this to be found especially in areas of quality and safety of food and beverage products, and there are quite a large number of quality, safety, and public health standards, both for content and formulation of products as well as their manufacturing, storage, and distribution. None of these areas touch the heart of food and beverage products and their creative and innovative side. There is no modularity or platform thinking to be found in food companies that would go beyond their own walls to create and innovate together with other companies in the industry and ultimately integrate retailers all for the ultimate benefit of consumers. Let me be more specific on this suggestion.

Today, every food company develops most of their products and even services largely on their "own," despite the growing trend of open innovation. Interestingly enough, one can even observe a push-back to this trend, especially from the "long timers" in the food industry who believe that holding on to as many elements of the value chain is the best way of generating value and profits. Although this is an understandable position because it pretty much reflects the stance of the traditional food industry, it denies change, even the slightest change. And the slightest change would simply be to open up to a vast amount of experts and expertise to be found outside the walls of the own company and ultimately let the "world" in.

Yes, I do realize that this frightens a lot of the experienced and well-aged managers of food companies because opening up, and in a sense letting in the world, goes hand in hand with losing control over something that always was controllable and that used to be successful for a long time. Now, let's add to this opening up the dimension of being driven by the R&D organization of the company, why not by all the R&D organizations in the entire industry, and now you can imagine that panic will set in. Well, before panic comes denial, refusal, and denigration of the new ideas: too ridiculous, not well enough thought through, not enough business insight, marketing knows better, and so on. Some of the remarks may have merit to them, but most of these and similar ones are almost solely based on fear and fear of loss of control. And there is more. In addition to these two elements of change, opening up and being driven or even steered by the R&D organization of the company, there is the suggestion to introduce the "customer (retailer) is king" as a new business approach, at least for those who have not adopted this yet. So, three novelties to digest in one go:

- Open innovation
- R&D sets the tone and steers the company
- Distributors are the real targets for the food company and should become real partners

Much has been written on open innovation, most prominent probably Henry Chesbrough in *Open Innovation* (2003). Needless to say that there is much more reading material on this topic, and I have personally practiced open innovation and written about it quite extensively (Traitler, Coleman, & Hofmann 2014).

10.1.3 Open innovation still remains much of a lip service approach

Chesbrough's groundbreaking publication is, while I write this, more than 13 years old and given the apparent merits and multiple successes of the open innovation approach, it is surprising to see that the uptake in the food industry, especially in large food companies, is disappointing to say the least. Yes, there is much lip service but when push comes to shove, when real commitment to the approach is needed, reality looks different and excuses are given why a certain collaborative project in open innovation mode was not pursued. It's a real shame. I know what I am writing about here because I was in the middle of this and had put a lot of energy behind establishing the approach of open innovation in the "disguise" of Innovation Partnerships in my former company. When I had left after about a four-year battle, my successor assured me that running these innovation partnerships would be his dream job. In reality after a month or so into his job, he lost all interest because he thought the continuation of the battle wouldn't be worthwhile. Yes, there was and still is much lip service. It comes back to what I discussed before: fear of loss of control and fear of loss of possibly great inherent value in the totally internally occupied value chain.

On the other hand, the branded-products food industry should be able to read the writing on the wall: retailers will gain even more power and are increasingly at the center of the action when it comes to connecting R&D and manufacturing with reaching out to consumers. If they (the retailers) believe or sense that the manufacturers are not dancing to their tunes, they will step up their efforts and copy branded products and make their own private label range of products. Yes, the brand owners pride themselves that their brands are so loved by the consumers, have fantastic quality, and represent a glorious lifestyle experience, but the reality is that the consumers of food products don't think in the same terms like they would purchase a car. Cars have a much stronger emotional connections and a certain status to them, which can probably never be matched by food products. And there are a few examples of food and beverage products that almost came to a mythical status, representing certain lifestyles of those who consume them. One example is Red Bull® drink, others could be M&M®, KitKat®, and Milo®, the latter especially in Southeast Asia. There are certainly a few other branded products that fit into this lifestyle-enhancing category, and I will let you decide, which ones you would add from your own perception and experience.

10.1.4 Brand strength is volatile

As I have just mentioned, retailers have an increasing lust to make their own store brands, and contrary to the apparently dominating general wisdom of avoiding vertical integration as much as possible, one can see an increasing number of store brands in most of the supermarkets. Some of these retailers, for instance the Migros Cooperative of Switzerland, started as the big copycat ever since Gottlieb Duttweiler founded it in Zürich in 1925. They would go shopping to other retailers and without any hesitation and in the shortest possible amount of time would have a copy product on their shelf. They still do this and have been extremely successful. One of their most comic copycat products was a decaffeinated coffee, which they called *Café Zaun*. You need to know that the original product from Germany was called *Kaffee HAG*. The (Swiss) German word *hag* means "fence" in English and so does *zaun*. Migros is not a small player and generated an annual revenue of CHF 27.3 billion in 2014; this amounts to approximately $28 billion, and this for a country of approximately 7.5 million people. Most of this revenue is generated in Switzerland. The success of their business model is quite remarkable and is largely built on this steadfast private label, store brand focus together with a strong emotional connection with their customers. And there are other store brand models such as Aldi Nord in Germany and many other countries worldwide or Walmart (mostly in the United States), just to name a few. Aldi is an interesting example because it started by accepting many branded food and beverage products in the beginning of their retail history. I do remember the struggles of my former company when we were feeling the pressure to be kicked from their shelves, especially our confectionary products, back in the late 1990s.

10.1.5 Store brands become more popular, or so it seems

Aldi Nord, by the way is the founder of the Albrecht family-owned Markus Foundation. This foundation owns 60 percent of Aldi Nord and since 1979 100 percent of Trader Joe's. Trader Joe's started as a small retail business back in 1958 under the name of Pronto Markets on Mission Street in South Pasadena, CA. It changed its name to Trader Joe's in 1967, and its original business model was different from what it is now. It really started by buying up the odd lots of shipments of top class French wine from a producer or trader who had gone bankrupt and sell this lot as long as the supply last. It is similar to Odd Lots in rural Ohio, later acquired by Big Lots. Once it was gone you could be sure that you didn't get more of this product. You had to wait for a new opportunity but you could get fantastic deals. Little by little, Aldi Nord made its mark and imposed the business model of restricted yet-still-large number of store brand products, all made for Trader Joe's, all this in a relaxed and customer friendly atmosphere. Store workers wear Aloha shirts and would typically engage in a conversation at the cash register. The only segment that is carrying almost exclusively branded products is wine, although even there Trader Joe's pushes the sale of cheap commodity wine, pretty much custom made for them like the Charles Shaw brand. This brand started in the early 2000s with bottles at a price point of $1.99. This price has not changed much over the years and has almost become a Trader Joe's store brand.

I tell these stories because they prove my point that, if the model is well thought through and even better executed, private label and store brands will inevitably become more popular and consumers will turn more frequently to them. So what is the answer to this trend that companies that develop and manufacture branded food or beverage products have to find? Well, I gave three possible scenarios, which I believe are part of the overall answer, and which by no means are either a silver bullet or the only answers.

Let me briefly repeat my initial answers.

- Increasingly turn to open innovation including retailers
- Have R&D set the tone for new product development and steer the company much more forcefully
- Recognize that distributors and retailers are the real targets and partners of all the efforts of any food company

10.1.6 Let's join forces

And yes, there is more that needs to be done and that can be done, provided that the company embraces change and is willing to go for it. Let me expand on this in the following. As far as the battle between branded products and private label or store brands is concerned I would apply the following. The first point is to accept the meaning of the simple saying: you are your best competitor, or in other words, you are the owner of a great brand and another company, your competition, copies your product successfully so that you fear for your market share or even worse, survival of your own brand. What to do? There are many answers to this dilemma such as putting more product fixed marketing expenses (PFME) behind your brand, renovating the package, generate a new flavor variant, lower the price, running a special add campaign with celebrity endorsement, and maybe a few more. I strongly believe that neither of these is the right answer, not even the combination of several of them is the right answer. The right approach is to recognize and accept the competitor, the store brand owner, as potential partner and possibly even offer the available manufacturing assets to produce for the competitor.

I do realize that this may raise quite a few eyebrows: you cannot do this, this is in conflict with antitrust laws, you give away too much of your own know-how, and other not constructive comments. And in part or each individual comment in today's development and manufacturing environment may have some merit. I look at the opportunities from a different angle, which is the new approach to R&D, especially how IP may be generated, owned, and managed. It is not efficient when one just thinks of how many overlapping efforts are done in the many R&D organizations of the various different food companies. Just imagine how much more efficiently this could be organized by joining forces and developing new products, processes, and services in the food industry in more sharing and almost precompetitive ways.

10.1.7 We have to accept that there are problems out there

This sounds maybe too radical but I believe that, especially given today's economic situation and budget consciousness that most companies demonstrate, the collaborative approach is the one that will have to be chosen, and it will have to be chosen in smart and profitable ways. Let me analyze and discuss this in more detail. Change is never easy and more radical change even less so. And that's what it's all about: radical change; some might even call this *revolution*. I am not a revolutionary and my suggestions and subsequently called-for desirable actions are partly based on long years of observations and partly on common sense, or what I prefer to call: being able to read and understand the writing on the wall.

One of the first steps in this change process is recognition and acceptance of the fact that today's situation in the food industry, especially for branded food manufacturers, is not a comfortable one, to put it nicely. For some, the situation is outright dramatic, for others less so yet they begin to see some hardship coming along their way, too. I do understand that this is a tricky terrain to talk about and food companies, especially the large ones, do not really want to tackle such writing on the wall openly; some discuss the issues, others prefer to sit them out. The right way, however, is to tackle issues head on and analyze best ways forward. So, what is this writing on the wall? Let me list a few that I strongly believe are clear to see and yet not enough dealt with by most of the food companies. And R&D plays a special role in this.

10.1.8 We need to take the consumers' fears seriously

First, there is a certain fatigue with consumers when it comes to "industrial food." Although most industrial food has advantages such as convenience, easy and fast to prepare, and known and expected price points for more efficient and foreseeable budgeting, consumers increasingly shed doubts on industrially manufactured food products in general. Such doubts include lack of naturalness, potentially unhealthy ingredients, possibly genetically modified agricultural raw materials, and more, rather diffuse and not well-articulated "angsts." Coming from a scientific background myself, I do not share the diffuse fears, but there is no choice and one has to take them seriously. That's my point here: I believe that most companies don't react to the consumers' fears in the most appropriate ways. Let me give one example: genetically modified organisms (or GMOs) are increasingly suspected by consumers to be unhealthy for them as well as for the environment where plants are grown as well as for the diversity in the surrounding habitats. Moreover, there are suspicions by the consumers that GMOs are only good for the agricultural industry, bringing them additional profits from hybrid seeds as well as tailor-made and fitting pesticides.

Again coming from a scientific background myself I would personally not share all of these apparent criticisms but that is of no relevance. The important point here is to recognize and acknowledge these consumer fears and respond in emphatic and not antagonistic ways. Some 15 or so years ago my former company hired a full-time expert with a whole network of other experts around him to basically fight the battle of the agriculture company that was (and still is) at the origin of the whole debate. Most of us didn't understand why we would fight their battle, but fight we did. And there was no way to win with most of the consumers, neither with legislation nor with most of the players in this entire mine field. I am not saying that we should not fight for what we believe is right, but this doesn't mean that we have to take on the entire world. Let us understand the consumers by almost entering into their heads—pretty much the same idea as in the movie *Being John Malkovich*—and look to the world through the eyes of a consumer and live their desires, dreams, hopes, and fears, justified or not, and empathize!

If a majority of consumers believe that a certain type of product or technology is not good for them, based on popular belief then maybe the food industry and especially the manufacturing companies should listen to this and not try to out-argue the arguments that are floating around, back and forth and in favor or against said new technology and product. One does not argue with consumers by slamming volumes of scientific literature on their breakfast table and hoping to convince them by such "overwhelming scientific evidence, proving that all is in order." It is not in order, even if the evidence would suggest so. And there is of course the other side: consumer groups and NGOs, and they are not on the side of the shining knights' army either. One has to understand that all such groups depend on raising funds from the public at large and the more spectacular their arguments and actions, the greater the likelihood that they will be heard and given. This may sound cynical, but most of these large and well-known activists' groups are in the business of fund-raising. I am not suggesting that individuals who fight for certain causes on behalf of such groups do it for money; no, I truly believe that they do it for the cause but the organization behind them is acting differently.

10.1.9 It's so confusing out there, please help me!

Having said this does not necessarily mean that their cause is unjust, but maybe their actions are often blown out of proportion and largely exaggerated simply to attract public attention. And the consumers are in the middle of companies, who try to sell their products entirely based on business goals, and NGOs, who attempt to go against such companies by putting out potentially flawed or skewed counterarguments surrounded by spectacular actions. I am a consumer, too, and how should I react to this? I am confused, despite the fact that I believe to be an "educated

THE MOST IMPORTANT FEARS OF CONSUMERS

- Not clean, unhealthy ingredients
- Long transportation distances
- Large carbon dioxide footprint (especially packaging)
- GMO—only good for industry
- Not sustainable
- Too much waste
- Too much speculation with raw materials
- Scarcity of water
- Climate change impacting agriculture

Figure 10.1 Consumers' fears.

consumer"; I continue to be confused and I start to become frustrated. Who's right and who's wrong? Is there a right or a wrong? I don't know, so I am just looing interest, especially in these debates about industrial food, GMOs, food safety, health concerns, sustainability, survival of agriculture and farmers, feeding the increasing world population, large heaps of food waste, speculation on agricultural raw materials, scarcity of water, global warming and…everything starts to spin in my head, and I am even more confused. Maybe I shouldn't consume industrial food products anymore and go back to the roots of food. I decide to buy my food increasingly at my local farmers' market, at the closest kama'aina market, or even better, directly at the farm. By doing this I would at least know where my food comes from and would become less confused and may calm down and sleep better.

Can I really? Where does the local farmer purchase his or her fertilizers, pesticides, or other necessary ingredients to grow food? Where does the food for his or her cows, pigs, chickens, or other animals come from? Most of the time it's probably not locally sourced, let alone grown. It's really complicated and I start to realize that this may not be the solution either. HELP! Can someone help me please? The answer is not a simple one, but should probably be; yes, however it's not going to be easy because it requires quite some dramatic and substantial changes on both sides. My topic here is not to incite and propose change to consumer groups or NGOs but rather to the industry. This is where change has to start and it has to be rather radical!

Consumers have fears, too; Figure 10.1 lists the most important ones. The list is by no means complete and you may want to complement it based on your personal experience.

10.1.10 The new business model 2.0

The first step of this new approach that I would call "new business model 2.0," which I had briefly described previously was:

- Increasingly turn to open innovation, additionally reaching out to retailers and include them in the R&D part of development of new products and services.
- Accept the fact that R&D is the real trend and tone-setter in your company and accept their guiding and steering role.

- Recognize and accept the maybe "unpleasant truth" that retailers are the real targets of all efforts of your company and should be considered partners much more than they already might be.

This represents major changes and is far from easy to be accepted by those in power in your company today; it would simply mean the end of the traditional powers and the beginning of a new distribution of roles in the company. That's not an easy feat, maybe even close to impossible, but it will have to be attempted if food companies want to remain successful on the long run. Accepting the three principles (which, by the way are only the first step of business model 2.0) requires some rethinking and reorganizing of the company and I would suggest the following generic structure and distribution as well as possible outsourcing of the different tasks. But let me first reiterate what I had written in another book (Traitler et al., 2014) about possible different company structures, each of which could be a perfect fit for food companies.

The product-centric company

Functions inside the company	*Functions outside the company*
→ **Office of Strategy**: Decides on the major direction(s) of the company; includes R&D and product and packaging development.	Finances, Legal
→ **Office of Design**: Decides on product nature, its "soul," branding and brand essence; is at the center of the product.	Advertising, Communication
→ **Office of Execution**: Has the task to "make" the product/packaging mix; includes Quality & Safety (Q&S)	Procurement, Logistics and Supply Chain
→ **Office of Marketing and Sales**: Has the responsibility to market, move, and sell the product	Regulatory

The design-centric company

Functions inside the company	*Functions outside the company*
→ **Office of Design Vision**: Decides on the major design and imagery of the company's products and packages	Finances
→ **Office of Design-Driven Product and Packaging Development**: Drives and manages R&D, packaging development.	Legal
→ **Office of Procurement**: Procuring raw and packaging materials is largely performed on a mix of design and cost drivers	Quality and Safety, Compliance
→ **Office of Manufacturing**: Manufacturing driven by the initially decided design principles, also influencing best manufacturing practices	Regulatory
→ **Office of Marketing and Sales**: Design-driven marketing and sales; also including communication and advertising	Logistics and Supply Chain, excluding Procurement

The consumer-, trade- and operator-centric company

Functions inside the company	*Functions outside the company*
→ **Office of Marketing, Communications and Advertising**: Decides on the major brand-related marketing, sales and advertising campaigns of the company's products and packages	Manufacturing and Operations
→ **Office of Design**: Decides on product nature, its "soul," branding and brand essence.	R&D
→ **Office of Supply Chain, Q&S, Distribution and Sales**: Logistics is the major product quality and cost drives	Product and Packaging Development
→ **Office of Regulatory Affairs**: Compliance with country and region specific regulatory requirements	Procurement
→ **Office of Legal & Finances**: Driving the costs down and managing liability risks	

The shareholder-centric company

Functions inside the company	*Functions outside the company*
→ **Office of Marketing & Sales**: Decides on the major brand related marketing and sales campaigns of the company's products and packages	R&D, Packaging Development
→ **Office of Procurement**: Procuring raw and packaging materials is largely performed on a mix of design and cost drivers	Manufacturing, Supply Chain
→ **Office of Communication & Advertisement**: Brand relevant communication and efficient advertising become major strategic drivers	Quality and Safety, Regulatory, Legal, Finances

I propose yet another setup for a food company, based on business model 2.0 and the suggestion that R&D steers: external resources are used and retailers become coworkers and target at the same time.

10.1.11 The R&D-centric company model 2.0 (equally applicable to model 2.1)

Functions inside the company	*Functions outside the company*
→ **Office of R&D Vision**: Develops the major future R&D directions and focus areas, thereby directing company's product, process, and services development → the company's sustainable future	Finances
→ **Office of R&D-Driven Product and Packaging Development**: Drives and manages packaging development including new materials and technologies in this area.	Legal

Functions inside the company	*Functions outside the company*
→ **Office of Design Vision and Execution**: Ensures a solid design basis and understanding for product, process, services and packaging development	Procurement, Logistics, Supply Chain
→ **Office of Manufacturing, Q&S, Compliance and Regulatory**: Manufacturing driven by the initially decided R&D and design vision and principles, especially influencing and driving best manufacturing practices	Advertising
→ **Office of Marketing and Sales**: R&D- and design-driven marketing and sales; extensive and close collaboration with retailers	Retailers: targets and partners Know-how providers: Suppliers, Universities, others (Open Innovation and Partnerships)

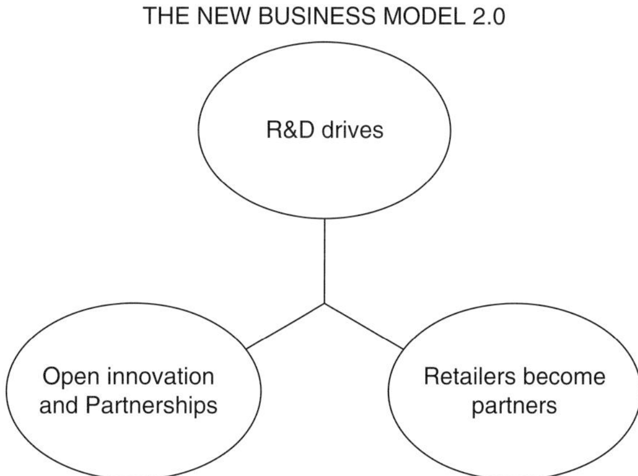

Figure 10.2 Suggested business model 2.0.

The R&D-centric model has at its core the belief that the future of a food company cannot be determined by finances and marketing, at least not alone and that a substantial and radical change is required to ascertain the sustainable future of any food company in this mine field of consumer confusion and NGO pressure.

Figure 10.2 illustrates the suggested new business model 2.0.

10.2 WHAT IF I HAD A MAGIC WAND? MY SECOND SET OF MAGIC TRICKS

As already stated, my first magic trick, if I had the power to perform one, was to create business model 2.0 and apply it to food companies, small or large. My second set of tricks would go farther, much farther I dare say, leading to business model 2.1. In simple words: I want to shift the food companies' focus away from developing, manufacturing, and selling industrial food products through supermarkets to consumers to developing, harnessing, and ultimately selling nutrition and health knowledge to the public at large, to consumers, to governments and NGOs, to health-related professions and professionals, in general to everyone who needs and relies on such knowledge. The emphasis is on selling know-how. This clearly is a tremendously large step away from today's industry focus. Today's focus and the industry's business model is approximately 1,000 years old, and it's probably about time to change or at least renew it. At the very least, it's time to talk about change and discuss various options.

Business model 2.1 is a possible logical extension of 2.0 and that's what I shall analyze and discuss in this section. It's not going to be an easy task and many in the present management establishment of food companies will most likely denigrate and ridicule what is being proposed; nevertheless, it needs to be looked at. I do not pretend that models 2.0 and 2.1 are the only models or the exact right and fitting for your specific company. However, I am convinced, based on many years of observation and personal experience, these two proposals are probably the best shots to guarantee a profitable and sustainable future to industrial food products and to the companies, which stand behind these products.

10.2.1 Change is inevitable in all areas!

I am convinced that despite all changes that will eventually come, there will still be industrial food products, but I am also convinced that the nature, composition, and presentation of these products will no longer be the same; they will inevitably have to be different because health- and nutrition-related new findings and changes in consumer perceptions and preferences will change. Even more so, retailers' business models will also change because consumers will increasingly shop online and will shop for a new type of products or "product modules." My prediction is that even the local farmers' markets will increasingly sell their products online, still keeping a large portion of their business on their close by, approachable, and emotionally engaging market stands.

Distribution and outgoing supply chains are changing dramatically as I write this. I just ordered items at Amazon.com on a Sunday around lunchtime and I had the merchandise at my door a few hours later, Sunday evening around 8 PM. Just imagine what that means for distribution of food and beverage products, and what it ultimately means for the food retailers! Safeway, Kroger, Tesco, Asda, or Walmart, just to name a few will actually become the new shipping companies and will look more like UPS, FedEx, or any other large shipper. Does that sound unbelievable? I don't think so, despite the still fairly low numbers of share of online business for the large food retailers.

Let me quote the following from an article in *The Guardian* that reflects the situation in the United Kingdom:

> *Grocers are cutting back on megastores, but rising internet food sales will see supermarkets sign up for twice as much online warehouse space this year.*
>
> *The major supermarkets—including Tesco, Asda and Waitrose—will this year commit to doubling the space devoted to [I]nternet distribution centres, known as dark stores, according to property agent Jones Lang LaSalle. Around 1.8 m sq ft of warehouse space is devoted to dark stores, but that is set to increase as online shopping transforms the retail sector.*

The internet is set to become a key battleground for the grocers in 2014 as Morrisons finally launches its online service in partnership with Ocado on Friday and rival supermarkets vie to entice shoppers with services such as same-day delivery and convenient "click and collect" locations. Morrisons and Tesco were among the retailers who declared an end to the "race for space" last year after admitting that out-size stores had been caught out by the emergence of online shopping and smaller convenience stores.

Grocers' online services are thought to have played an important part in the tussle for customers over Christmas. As much as 15% of UK grocery sales, worth £900m, are anticipated to have been booked online in the four days from 20-23 December, according to analysts at Verdict, compared with an average of 5.5% throughout 2013. Verdict is predicting that 6% of grocery sales will be made online this year while the market is set to double in value over the next five years to £13bn, according to IGD, the grocery market research body.

The importance of internet stores for non-food retailers this Christmas has been underlined by John Lewis and Next, which both reported double digit rises in online stores over the festive season.

While the proportion of groceries sold online is much smaller than other sectors, the vast size of the food market means it is a key avenue of growth for under-pressure mainline supermarkets. Stuart Rose, chairman of Ocado, told the Guardian that the online grocery market had reached the "point of inflection." "The future for online food retail is very exciting. We are at the beginning of a revolution and the pace of change is accelerating. It's online showtime," he said.

10.2.2 The new product will be know-how

Changes in the ways that retailers sell their products—actually the branded and store brand products that are manufactured by others—will inevitably dramatically change the ways such food products will be developed, manufactured, and brought to the consumers. Business model 2.0 only takes care of one important aspect of this change, namely the recognition of the new roles that retailers will play. It does not yet recognize the need to fundamentally change the product that is ultimately sold. Model 2.1 suggests that the new product is know-how. It's the same transition that one can see from brick-and-mortar locations that will ultimately be downsized to a much greater online business. To successfully sell know-how, food companies have to do quite a large number of changes in their business approach, content, people, and setup. The idea here is first to:

- Recognize that change is inevitable
- Recognize that such change will impact the company's business model
- Recognize that change will have to be implemented gradually, ideally running models 2.0 and 2.1 in parallel for still quite some time
- Recognize that to respond to all aspects of change the company has to be reorganized: its setup, its content, its people, and its ways of executing and selling the new product know-how

And there are probably a few more recognitions required to become successful in this new environment. Let me reiterate that I strongly believe that the only branch of the company who has the ability and (hopefully) the guts to get such changes under way is its R&D organization. By nature, almost everyone in the R&D group is curious, intends to find something new and exciting all the time, and pushes the boundaries. So, my suggestion is to mandate R&D in the company to push change through the organization as much as they can. And it goes without saying that they cannot do this alone and they need everyone else in the company to help them achieve the demanding goals. However, it should be guaranteed that R&D is empowered to go ahead in the new direction and pull the entire company along with them.

10.2.3 That's what's important for business model 2.1

One of the prerequisites for 2.1 to work is the acceptance of 2.0, at least in part: Recognize that distributors and retailers are the real targets and partners of all the efforts of any food company.

This is the real first step that is necessary to successfully progressing toward business model 2.1. So, what's in 2.1? Here's my first list:

- Building up and gradual transition to selling proprietary know-how through different channels such as experts whose work place is off site or on Internet platforms and professional Internet-based apps.
- Based on the strengthened relationships with retailers use both their brick-and-mortar locations and Internet platforms jointly for activities of selling nutritional and health and well-being related know-how.
- Grow and promote understanding and acceptance of consumers for the new "your company name inside" approach, similar to what the computer industry perfected many years ago and what can increasingly be seen in areas such as digital photography emphasizing the usage of CMOS chips.
- Adapt the hiring approach of your company and increasingly hire personnel with health and nutrition expertise combined with excellent communication skills, which will become the new marketing.
- Train your existing personnel accordingly and in preparation for the new business model 2.1.
- Most importantly make personal and personalized expert advice the core of your business both, in person as well as online.
- Include "chat-with-the-experts" platforms, ideally based on loyalty programs similar to the ones in the airline industry and through membership and long-term subscriptions.

This list describing the major elements necessary to advance the business model 2.1 is by no means complete, and like all other existing business models in any industry, it never will be complete but grow and evolve as the company grows and evolves. However, one overriding principle becomes apparent from this list: because it is knowledge based and know-how driven, the R&D organization is in the driver seat and has the task of taking the company forward to the expected successes that can be achieved through this new business model. It's not an easy undertaking because it's radical and in part has been tested in the industry in the past and was

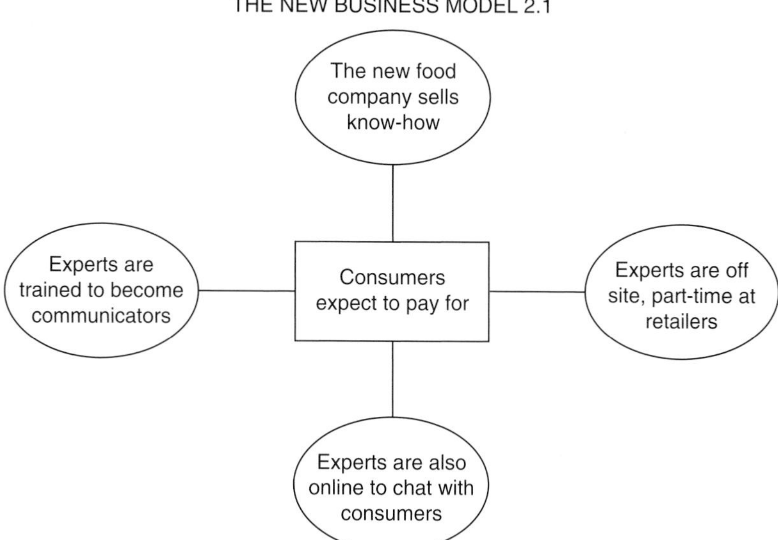

Figure 10.3 Suggested business model 2.1.

not successful. In my opinion it failed because this model was still embedded in the old company ways, in the traditional structures of a business model that is almost one thousand years old, if not older. I refer here to the case of Jenny Craig that was acquired by the Nestlé Company in June 2006, and which, after a few years of struggle to make it fit into the existing Nestlé business model was again sold off in October 2013. It simply didn't work in the present model but that is not to say that it does not work outside the traditional food company framework. I shall discuss this case in more detail in the following chapter.

Figure 10.3 illustrates the suggested new business model 2.1.

10.2.4 Here are the details

Let me analyze and discuss the various elements for success in more detail. The first one in the list is making experts in your company the future sales agents of their (the company's!) know-how. The experts are part of the R&D organization and their workspace is gradually changing over time as they go out and interact with retailers and ultimately consumers. They are increasingly found out in the field at the retailers' locations. Complementary to being on location, the experts can become agents to answer questions and concerns of consumers who are paying for such advice. I am confident that people are willing to pay for such a service because uncertainty about nutritional values, health impact, and potential negative side effects of ingredients in industrial food are increasing. This is not to say that food necessarily becomes less safe, but confusion of the public linked to food is on the rise and consumers will inevitably have more questions, which cannot simply be answered by the CEO or the PR department of the company. It's really the task of the experts to do this job, and this will require new skills and talents, different from the traditional scientists or engineers diligently working in the lab or on the factory floor and only being concerned with the technical and innovation well-being of the company. A new "breed" especially of scientists is required!

I have listed two major prerequisites for this to successfully happen in the new business model 2.1: Retailers have to accept the new proximity of food companies as laid out in the discussion of business model 2.0, and the consumers need to grow in this new environment, this new value that they can gain (i.e., purchase) from a food company. This new value is knowledge, knowledge about "my food," its composition, its nutritional value, its potential health implications—both positive as well as negative—and other elements such as value for money, sustainability of food, and food products in general, reducing waste levels by smart shopping, preparing and preserving, and probably a few more. Many of these elements are important topics for any community at large and governments of any country and of any kind (federal, regional, communal, or any other setup) have to have a vested public health conscience and interest in cutting public spending for health-related issues. Additionally, the entire healthcare industry is mostly interested in spending as little as possible on the insured to maximize or simply optimize their profit margins. All these organizations and groups are becoming natural financial supporters and partners for the educational aspect of business model 2.1 and need to be brought into this model from the beginning. For instance, health insurance companies, which already contribute for instance to personal fitness programs, can equally be supporters of good food and nutrition related educational activities as proposed in model 2.1.

To bring retailers into this model more easily, they have to become part of this entire group of know-how transfer, and their already traditional role as product and service providers to consumers is helpful. Additionally, they have to profit from any financial support of any of the third parties such as governments, health insurers, NGOs, and foundations that are supporting public health at large. I do realize that when it comes to money, people and organizations become short-sighted and think of themselves first and foremost and easily lose sight of the long-term effects and benefits that come from such a strategy (i.e., to dramatically improve nutritional and health status of large parts of the population and thereby reducing public and private health spending in equally important ways). The expected savings are split and the new food company, the one that operates with business model 2.1 earns its money directly from the consumers for the advice they receive and from a prenegotiated amount of reduced taxes from the public sector, as well as participation in profits coming from the private sector. Insurance companies and food companies become partners early on; food companies almost become an extended arm of healthcare insurers and share the expected increased profits. Sounds to good to be true? Maybe, but nothing that is radical is easy to achieve and should not be dismissed until it is solidly put into practice and tried out in professional ways.

10.2.5 Some calculations, just examples

Estimates of investments versus expected overall returns for the various players in this model 2.1 are the most crucial element in this entire equation and it will not be an easy task. Although it is fairly easy to calculate the investments—or costs, as some may want to call it—that any given food company would have to bear by simply determining how many experts would use which percentage of their productive time to cover a given, desirable size of consumer segment in any given market, maybe one or two to begin with. It goes without saying that one would kick model 2.1 off in carefully crafted first steps and with a limited time investment from required experts. Let me give a simple example: if a globally active food company starts in a place like Singapore, it would have a wonderfully compact, rather homogenous and affluent test market with a well-defined population mix as well as mix of retailers. If the same food company would collaborate with two different supermarket chains in 50 of their local outlets and would use a total pool of 60 experts at one-third of their time to be present in the different locations to interact with consumers and this on fixed days, for example three times a week, the total investment would be 20 full-time equivalents (FTEs) at an average of US$ 350,000 per FTE (i.e., $7 million per year). If the same company would use another third of the experts' time to participate in chats with consumers, we would add another $7 million for a total of $14 million.

10.2.6 The company can earn more with model 2.1!

If the food company makes an estimated annual revenue of $1 billion and an annual profit of, for instance, $140 million, the additional investment into the experts' time amounts to 10 percent of profit. Wow, we cannot accept this, many of you would say. My prediction, however is that additional profits that these experts can generate for the company by far outweigh the additional costs, maybe not in year one but, like for most new products in that industry at least at the end of year 3 of the inception of such a model. Out of Singapore's total population of approximately 5.6 million people (numbers for 2015), it would only require 3.5 percent or 200,000 who would become members of such a "chat/interact with the food experts" program, paying $100 per year; this would give them total access to nutritional and food related advice, which potentially can be offset by a tax relief or an insurer contribution (much like the insurer pays for part of gym membership costs). This would mean a zero additional cost for consumers in such a program based on model 2.1. The expected additional income for the company in this example would amount to $20 million, leading to an additional profit of approximately $6 million (based on the estimated costs of $14 million); not bad and simply an improvement of the margin by some 4.3 percent or 430 basis points. This looks quite a dramatic improvement, achieved by a simple and careful evolutionary change of the existing business model without dismantling anything of the old model (yet)! The best part of this is that it might not even be necessary to hire and additional 40 FTEs (remember: the example called for two-thirds of the work hours of 60 people). To begin with, existing R&D resources should be used more efficiently by re-dimensioning ongoing R&D programs and by increasingly using open innovation and innovation partnerships. The company's experts would undergo a tremendously efficient field training by speaking directly to consumers—either in person or through Internet chats—and over time would become so much more valuable for the company. There are only advantages to such a business model 2.1. The only difficulty still remains: the "old brains," which know it all, did it all, and are resistant to change, any change that is. I do hope that the arguments and examples can serve as a solid basis for further in-depth debates in many food companies that see their own future in a different light than the one that is shining today. At least, the arguments should be a basis for critical discussions and refinement of numbers in real-life scenarios.

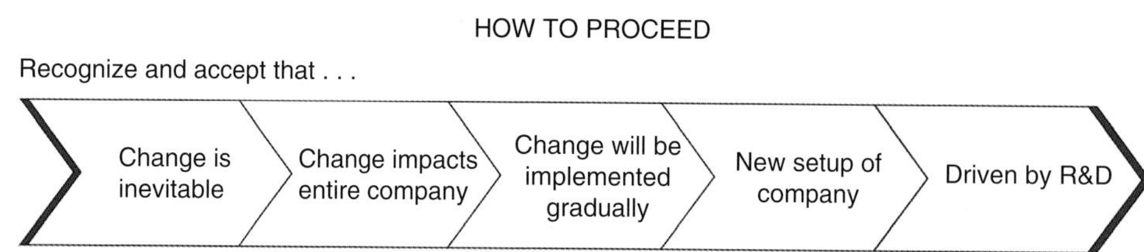

Figure 10.4 Recognize and accept these necessary steps toward an R&D-driven organization.

You may come up with different numbers, maybe even better ones, who knows? It is definitely worth a try, and business model 2.1 may become a much bigger reality than today's company executives have ever dared to dream of.

10.2.7 More changes: A new type of employee

And that's just the beginning. There is more to come like gradually increasing the part of this service activity and at the same time potentially decreasing the lesser value of manufacturing simple food products, leaving that to co-manufacturers, still under the control of the food company. This means that quite some substantial changes will have to be initiated especially as far as hiring and training of expert personnel is concerned. It will be important to actually go one step further back and gain more influence to appropriately modify educational curricula at university level as well. Scientists have to have additional training in communication, especially communicating directly with consumers while being outside their comfort zone like a lab or a pilot plant. This is not an easy undertaking because it would mean that universities have to be prepared to make the necessary changes and add more course work to their already busy schedules. Taking communication courses—theory and learning in the field—should be compulsory at least during the student's doctoral thesis times and should be included in the final exams through appropriate tests, again including field tests. It should be like getting a driver license: you have to pass both a written test and a driving test. That's exactly what is proposed here. I shall discuss this in more detail in the following section.

Figure 10.4 illustrates suggested general steps how to proceed to achieve the goal of an R&D-driven organization.

10.3 THE NEW SCIENTISTS AND ENGINEERS: A NEW TYPE OF PEOPLE

It is fairly easy to see that both new business models 2.0 and 2.1 will require a new "breed" of scientists, and to some degree, also engineers. More outgoing and more eager to explain and sell their work and the exciting new findings that come with these. So, what has to change? In my experience the biggest hurdle for scientists to be heard, let alone understood, is the way they talk, the kind of words they use, and the inability to focus on simple wording and making their message understandable and graspable for a non-scientific person. I strongly believe that the new scientists have the ability to learn this. The question is, do they have the desire? In the "old school," scientists are taught to follow a specific scientific approach to problems, be curious, be meticulous, and be suspicious with regard to the easy solutions. Scientific protocol requires asking the right questions, coming up with the appropriate hypotheses, and proving or disproving these by performing the fitting experiments. Finally, it is almost built in the genes of scientists and engineers for that matter to never see anything as truly finished; there is never a real end point to their discovery work and if not they themselves but following generations of scientists or engineers will continue their work building on existing results and evidence from those before.

In principle it's a good thing not to believe that one has resolved all open scientific and technical problems, and it shows that scientists and engineers can be rather humble and accept the non-finite character of their work. However, when consumers have important pertinent questions with regard to nutrition and potential health impacts of the food they eat and the nutritionists tell them that "there is evidence pointing to the preliminary conclusion that there might be a correlation…," much confusion may arise leading to suspicion or even mistrust of the nutritional scientists. No wonder that so many opinions on good and healthy food are out there and are announced as the truth of the day. There may be a new truth tomorrow, and yet another one the day after. I am not arguing with the underlying fact that science is an evolutionary process in itself, but that is not helpful at all to the ordinary person, the typical consumer who has legitimate questions or even doubts. It needs a different kind of training for scientists and also the engineers to become sensitive to such questions and being capable of giving definitive answers, even when they are well aware that definitive means, "to the best of our present-day knowledge these are the answers we have."

10.3.1 The new educational focus: Communicate

This not only means that scientists accept to talk about today's knowledge in an understandable way but also in "definitive mode," thereby answering all possible relevant questions in reassuring and competent ways. The big change for them is to talk about results without having all the facts coming from findings in the future. It also means that they should not respond in ways that clearly indicate that they always want to leave an "escape path" open for them: "I have to emphasize that our findings are only preliminary and more in-depth research is still needed." As briefly mentioned in the previous section, this requires changes in education, probably best achieved through additional course and practical work during the graduate studies, be it for an engineering degree or a PhD. And such courses have to become compulsory, so no excuse possible that one wants to concentrate on "pure science," whatever that is!

Of course education begins with the definition of an appropriate curriculum, but more importantly with the right teachers, highly motivated to convey the best possible know-how combined with learning what can best be achieved through such knowledge. Much has been said and written about the most important part of any doctoral thesis work is to learn to work independently, recognize and solve problems in independent ways, and combine this recognition with analysis and ultimate synthesis leading to the most promising and reproducible results. You may want to add more specifics to this definition, but in my book it's this "independence" that tops it for me.

Independence has one big potential flaw, though; it leads you to believe that you don't need anyone else to progress and succeed, so you seemingly don't need to be really communicative and that's a big problem. Yes, scientists and engineers communicate by talking about their work, new findings and exciting discoveries to their colleagues at conferences and other technical gatherings, and they speak in a lingo that only their likes comprehend but certainly no one outside this inner circle of scientific and technical savants.

10.3.2 Choose your words and help me to understand

By no means do I suggest that those outside this inner circle would be limited in their intellectual capacity to comprehend complex matters, but the technical lingo is so different and uses so much tech slang that it's really difficult to grasp. So, if you are not in this inner circle you will lose interest quickly and not much what was explained to you in highly complex and almost enigmatic terms will actually stick and make an impact. And that's exactly what is needed: a complex message recounted in widely and easily understood words that stick with the receiving end, in my example, consumers and shoppers of food and beverage products. This has to become part of a new curriculum for graduate students, and faculties will have to expand what they are offering to more than just a few almost alibi courses on preparing them for the "world outside" when they get close to the end of their studies.

The importance of great communication, well understood by everyone cannot be stressed enough; without learning to master such great communication nothing will change and graduates coming out of universities maybe great in tech talk and may have even better jokes in their playbook but will not be ready to fill the necessary slots to make business model 2.1 the success that it is expected to become. An important part of adopting the idea of putting communication in the center of the new scientists' and engineers' work is the recognition of opening up one's view of the world around, which is much bigger than the lab or the pilot plant. It also means that the new ways will embrace a holistic view of new product, service, and process development by acquiring additional skills such as design—product and packaging—marketing, sales, and procurement, just to name a few. Depending on the specific product and service portfolio of the company this mix may be differing slightly and some of these skills can also be taught and perfected on the job.

10.3.3 That's what it takes

The new species of R&D people will be open minded, curious, humble, and willing to learn all the time; great communicators who like to tell great stories; willing and desirous to interact with just about anyone; and in simple words, they will have to become multitalented. On top of being experts in their own field they have to become communication experts. As it's never too late for anything in life, such expertise can and must be acquired by every R&D person working in the industry today by appropriate courses or internships that can and must be organized and offered by the company. Internships may be found in the media and Radio and TV industry, fund-raisers, hotel and hospitality industry (e.g. receptionist or tourist guide), team sports clubs, and probably many more, for you to discover. The point here is to realize that it's never too late and members of the existing personnel can be molded into the new realities of business model 2.1.

I am realistic enough not to believe that everyone can be remolded, but many will follow, and many more once first success stories from these new communication-centric activities come back from the field.

The real learning and conclusion to this topic of the new type of people required to run the show can be summarized simply:

- Universities have to adapt their offerings to technical graduate students and include teaching of communication skills as well as a few others such as basic notions of design, marketing, sales, and procurement (long term)
- Companies have to offer tailor made courses for continuous learning in addition to sending their R&D personnel to short external internships (short to medium term)
- Existing personnel will have to become motivated enough to follow such courses or internships by "actionable promises" with positive impact to their present and future careers (short to medium term)

It is certainly not an easy feat to convince the company's management to put any of this into place, let alone the organizational difficulties that come with these suggested changes. But then, no change is easy, why would this one be any different?

10.4 THE NEW R&D ORGANIZATION

The entire book and every chapter that led up to this one suggested the need to transform the R&D organization of the company to become the new, the real driver of the company. Now we need to discuss the ramifications that such a change inevitably brings about. Ramifications are just another word for change, and as has been said, change is difficult to accept for those who find themselves in the middle of this and is more difficult to actually put in action. The problem with change is that it has been abused so many times in the past for many initiatives that were not logical, justified, and only served a small group of people, typically those who are in power. This book is not a revolutionary playbook, it is simply a call for action to do the obvious and the necessary. I do realize that what may appear obvious and necessary to me does not look the same for others; however, I reserve the privilege for me that what I write about and what I discuss is lived experience and based on many years of observations and having been a "change-agent" myself. One could argue, rightly so, that I was not successful in changing much in my former company because I was not able to introduce many of the ideas and suggestions for change in my former company. And you maybe right, in part at least.

I do know, however that we—my colleagues and I—were able to change attitudes of some of the people in the R&D organization, although not enough; we were able to bring about some logical structural and organizational changes, although not enough; and we were able to initiate some changes to content and direction of the complex project work in the R&D organization, although by far not enough. I do admit that one could argue that we failed, but that's too easy an explanation. Changes in organizations of any size but especially the larger ones are really difficult to implement, and you will always find tons of naysayers who have lots of arguments to explain why nothing should be changed and everything should remain as it is. After all, "we are doing pretty well, so no need to change; it's too risky."

This is all too human, and it is difficult to find the right and convincing counterarguments. Much or rather most of what was written in this book thus far leads up to have such counter-arguments at hand to respond and convince the naysayers.

10.4.1 Change is a risky business

One of the reasons why more change has not been brought to the organization and setup of food companies, and especially their R&D groups, can simply be explained by fear—not so much fear of change but fear of consequences for individuals, especially new responsibilities that come with such change and not everyone is willing to accept and shoulder. It is more comfortable to remain in the present than to accept an unknown and potentially increased responsibility and accountability that comes with such a new order. We have discussed the new type of scientists and engineers, and the big hope is that such a new type, or rather a new generation, will have the guts to fight for and accept new and increased new responsibilities that inevitably will come with a heightened role that the new R&D organization shall play.

I had suggested a possible generic structure of a company that is led by the R&D group. I do realize, again that the idea that R&D would run a company, a food company that is, might frighten some of the establishment because it would take away their power and would change a business model that is approximately one thousand years old.

Anyhow, let's try it and go through some of the details. And please do remember that we apply the new business models—either 2.0 or both 2.0 and 2.1—and that we work with a new breed of scientists and engineers that actually have learned to communicate properly and have the guts to take on additional skills and responsibilities such as marketing, sales, and most of all are able to understandably talk to customers and consumers. Let's give it a try!

10.4.2 Here's the list

I suggest a series of changes from the present-day, traditional setup and approach a new, possibly radical and disruptive one that is best suited to fit the new business models 2.0 and 2.1:

- The first important yet still general change should reshape the R&D group away from being a follower to becoming a leader. Based on my experience, most of the R&D groups in food companies, smaller as well as larger, largely follow what they are instructed to do by the business or directly by management. Although many in the R&D would pretend to be independent thinkers and doers, this is far from the reality.
- The second modification has to happen in the personnel structure, away from a structure that is still hierarchically driven to one that is flat and fosters self-organization and increased taking of responsibilities. As discussed previously, this might be a rather frightful experience for some because taking on responsibilities and standing up for one's successes and failures is not unequivocally embraced by all.

- A further important transition is one in which fixed structures are discarded to the benefit of introducing flowing structures based on the real development needs of the company. Fixed structures, as they exist today basically everywhere have the great disadvantage of being slow, unresponsive, and heavy to carry through the organization, especially difficult if not impossible to change or dismantle unless people are let go. The new setup can have different faces, from a general expert pool available "on-the-go" to a modular setup or anything else to be further discussed based on the company's real needs.
- As mentioned, projects are at the heart of any activity in most companies, especially in food companies. Projects are defined, discussed, rediscussed, briefed, kicked-off, status-checked and rechecked, reoriented, and after much more ado, eventually terminated, at least tentatively. It's a heavy process and I suggest to change this and move away from a situation of exactly defined projects to one of creative trial and error. I do realize that this adds complexity—some may say impossibility—to a proper and predictable resourcing of such trial-and-error activities, but that's the whole point: to move away from a closed and asphyxiating system to an open and freely breathing one.
- The next point on my list is to change from a system in which the business or management defines content to one in which content is defined by the R&D organization. To some degree this may already be done in some or many companies today, but it's often content of "lesser importance," which is determined and defined by R&D. This should change to a situation of overall responsibility of which content makes up the workload given to R&D.
- Another important change is suggested in regard to budget and resource allocations and how this is organized in a company. For whatever reasons, including accounting considerations, the different functions in a company have defined budgets and resource allowances that are expressed in number of people or FTEs. This has to change to a situation in which content and workload as defined by R&D define the budget needs behind required resources, thereby leading to much-needed increased flexibility and especially speed of execution.
- A bit in the same direction as giving up fixed and heavy structures is the call for change from formal teams, departments, and workgroups to self-forming work groups that are defined based on direct needs. Such groups are able to respond much faster to any new request and are able to accomplish results more expediently.
- Finally, much in keeping in particular with the suggested new business model 2.0, the focus has to turn away from looking inside and mostly communicating internally to a more open look to the outside and communicating to the outside, directly joining forces with the customers (the retailers) and speaking to consumers and all this not filtered through the eyes and ears of marketing and sales but directly through the R&D organization.

I do admit that many of these suggestion represent rather radical changes, possibly difficult to accept and more difficult to implement and eventually live as the new normal. The following chapter analyzes and discusses the different models and scenarios in much detail.

Figure 10.5 depicts the suggested transformations and changes that will make every food company not only stronger but definitely more credible and successful.

THE TRANSFORMATION OF R&D

FROM ➡	TO
Following	Leading
Hierarchical	Flat
Fixed structures	Flowing organization
Defined projects	Creative trial and error
Content defined by business	Content defined and driven by R&D
FTE-based budget allowance	Content- and workload-determined budgets
Formal workgroups and teams	Self-forming workgroups defined by real needs
Inside-looking and communicating	Outward-looking and speaking to consumers

Figure 10.5 The transformations.

10.5 SUMMARY AND MAJOR LEARNING

This chapter set out to develop a new vision for the future of R&D in the food industry, a vision that has the ambition to be almost utopian and certainly daring. It's probably easy to criticize such an endeavor and I ask the reader to do this, however, based on the recognition of the flaws of the present situation, with the one question in mind: How would I go about reforming the company and especially its R&D group to make the whole organization perform optimally and sustainably?

The following describes the major talking points and learning from this chapter.

- The present-day situation of food companies is not as rosy as it may appear by looking purely at the numbers and annual reports. The mantra of every food company is still: "Consumer is king (or queen)"
- It was suggested that the first "magic trick" to make change happen would be to come to a recognition that the new king or queen is the customer, in other words the retailer, the supermarket, or any food and beverage seller; this should become the new mantra.
- It was suggested to break up the boundaries between manufacturer and retailer and achieve a new and intimate level of collaboration, beneficial for both parties and ultimately for the consumer.

- Elements to achieve this new quality of collaboration between food companies and retailers include embracing and applying open innovation, moreover that the R&D group sets the tone and steers the company as well as the acceptance of real change.
- Despite many success stories around open innovation food companies, the large ones see this rather as a threat and loss of control than an enrichment and pathway to increased success.
- Important reasons that forcefully justify an intensified collaboration with retailers and the increased application of open innovation and partnerships were mentioned and examples were discussed. It was suggested that even strong brands are easily being copied and attacked by retailers through store brands and private labels. Branded food companies have almost always refused to manufacture store brands. There is much room for improvement.
- Close development collaboration between food manufacturers and retailers can be challenging in view of cartel and antitrust regulations; however, the new approach to R&D described in this book shows ways how this can be overcome. These include joint developments in precompetitive ways, development of IP combined with profitable licensing, and increased usage of overlapping IP and know-how that is generated in parallel in many parts of the food industry.
- Industrial food is increasingly under attack by NGOs, consumer groups, and consumers; food manufacturers are under direct attack by governments, especially for possible wrongdoing. In today's setup and approach, most of the responses by the food industry, especially by food companies, is not appropriate and often counterproductive, potentially leading to more confusion and rejection.
- This chapter suggested, analyzed, and discussed two radical new business models for the food industry. It has to be recognized that the present business model of food companies is dating back around one thousand years (!) to the first beverage and food manufacturers and typically consists of: development, manufacture, and sale of products, possibly including services such as package and product convenience, easy availability, quality and safety, and a few others.
- The first new business model 2.0 suggested in this chapter consists of the elements that:
  - Increasingly turn to open innovation, additionally reaching out to retailers and include them already in the R&D part of new products and service development.
  - Turn to the new reality in which R&D is the real trend and tone-setter in the company and accept their guiding and steering role.
  - Recognize and accept the "unpleasant truth" that retailers are the real targets of all efforts undertaken by your company.
- Based on model 2.0, a generic new R&D-centric company model 2.0 was suggested and briefly discussed. This new company model has at its core the belief or rather the conviction that the future of any food company cannot be determined by finances and marketing alone.
- A second new business model was proposed, analyzed, and discussed. This model 2.1 builds on the recognition that industrial food products may become so commoditized and

might partly disappear that a new element in the value chain has to be established. Business model 2.1 suggests that know-how and expertise is this new element: the new product will be know-how. Model 2.1 builds on one of the core elements of model 2.0, namely the new proximity between manufacturer and retailer. Experts from the manufacturer can work off site at the retailers and directly interact with real people, the consumers.

- Major change elements in business model 2.1 are: gradually transition to selling proprietary know-how through different channels including both brick-and-mortar as well as the Internet. Have experts working off-site complementary to their on-site activities. Grow and promote understanding and acceptance with consumers. Understand that a new type of scientists and engineers is needed if this is to be successful. Hiring, training, and even university education has to be appropriately adapted.
- Some practical examples of external implementation of experts were discussed, and costs and rewards were calculated, demonstrating potentially rewarding results, adding importantly to present revenue and especially profits.
- The new type of employees that fit into business model 2.1 was extensively discussed and communication skills were identified as the major competences that the "new employee" absolutely needs to acquire and use, especially talking to consumers in understandable and captivating ways.
- The new R&D organization that fits the suggested business models was discussed in some detail, especially in light of the importance that the new company leader, the R&D group, will gain. This brings new responsibilities and accountabilities for these new leaders, and not everyone in the R&D organization may be up to this challenge.
- Finally, the importance of embracing and accepting change was discussed and the following transformations were listed: from following to leading and driving, from hierarchical structures to loose, self-forming structures, from fixed structures to flowing structures, from defined projects to creative trial-and-error activities, from business-defined to content-defined activities, from FTE-based budgets to workload requirements–linked budget allocations, from formal teams to self-forming work groups, and finally from looking inside and communicating internally to opening up and looking and communicating to the outside.

REFERENCES

Butler, S., 2014, January 6, "Grocers rush to open dark stores as online food shopping expands," retrieved from http://www.theguardian.com/business/2014/jan/06/supermarkets-open-dark-stores-online-food-shopping-expands

Chesbrough, H. 2003, *Open Innovation, The New Imperative for Creating and Profiting from Technology*, Harvard Business School Press, Boston.

Traitler, H., Coleman, B., and Hofmann, K. 2014, *Food Industry Design, Technology and Innovation* Wiley-Blackwell, Hoboken, NJ.

11 Testing the hypotheses

There's two possible outcomes: if the result confirms the hypothesis, then you've made a discovery. If the result is contrary to the hypothesis, then you've made a discovery.

Enrico Fermi

11.1 TOO GOOD TO BE TRUE OR SIMPLY WRONG?

You may have asked this question, especially after reading through the preceding chapter regarding "utopia" for a new R&D approach and organization in the food industry. You may argue, rightly so, that it is always easier to propose something than putting such proposal into practice. Yes, this is true, but it is equally true that proposing something that is pretty daringly different from today's situation is infinitely more difficult than proposing nothing. I have seen this time and again: people were complaining about a suboptimal situation in their environment and asked for change, but did neither think about how such change should present itself nor did they ever dare to propose change even if they had some vague or even good ideas. And yes, it is easy to suggest and have a possible solution at hand or have no solution at all (something that politicians do all the time) than to actually do something. The problem with "doing something" is that one could be wrong or go wrong and that's a risk. So, doing something is for risk-takers, just suggesting something is more for politicians.

The ideal situation, though, is of course to combine both: suggesting and doing. This book, and especially the preceding chapter is suggesting a lot, and it also gives solutions and pathways as to how the doing-part could look like. Being a written text, the real doing has to be left to the risk-takers, well you the readers, at least a good portion of them. This present chapter is all about critically questioning and analyzing the proposed pathways for change, rather radical change that is; not many readers may even like the idea of radical change in an organization such as a food company that is old and whose business model is principally 1,000 years old. Long-standing habits have carved out cozy and familiar pathways like water and wind in the canyons of mountain rivers. These are difficult to alter.

Food Industry R&D: A New Approach, First Edition. Helmut Traitler, Birgit Coleman and Adam Burbidge.

11.1.1 Let's look at business model 2.0 first

The previous chapter suggested, analyzed, and discussed two major new, or rather new, business models. I characterize it "rather new" because other nonfood industries have adopted some or the majority of aspects in these models already. They are definitely new and certainly disruptive for the vast majority of food companies. Let me go to some serious dissecting and challenging business model 2.0 first.

The three elements of model 2.0 are:

- Retailers become real partners and co-developers, the real extended arm of the food company.
- R&D is the driver in the company.
- Open innovation and partnerships with external expert providers are fully embraced, much more than at present.

11.1.2 Let me take stock

Are these really serious and practical proposals? Let me take the first point in which it is suggested that retailers become real partners and joint actors in development, a real extended arm of the food company, the one that has created and manufactured a large number of food and beverage products. What's today's typical status? Well, the climate between food manufacturers and retailers has become milder and apparently more collaborative. How come all retailers still charge listing fees for the manufactures' products just to put products on the retailers' shelves? It's like renting a hotel room and being charged twice: once for night's stay and a second time to be handed the key that opens the room. You wouldn't expect to be treated like this in a hotel, and yet, every food company accepts being charged twice: listing fees and a percentage of the revenue of the product that is sold from their shelf. To come back to my hotel comparison: you would probably mildly call the double charging a real "rip-off," but every food manufacturer quite naturally, although grudgingly, accepts such rip-off. How come? I have a personal list of answers to this question, which might not be totally in sync with the official line, but here they are:

- It's the system, there is no other way to get on the retailer's shelf (the obvious answer).
- My product is kind of weak, and the retailer helps me to convince the consumers to buy it anyhow.
- The retailer promises me a more prominent spot in their retail space; I am willing to pay a premium for this promise.
- The retailer intends to remove my product from the shelf, and I pay a bit more to keep it there for a bit longer. Consumers might still get to like the product; the intended de-listing was just too early.
- The retailer is really successful and a safe bet for great sales, so I am prepared to pay the requested listing fee.
- The tax write-off related to the listing fee is (almost) outweighing the loss of profits.

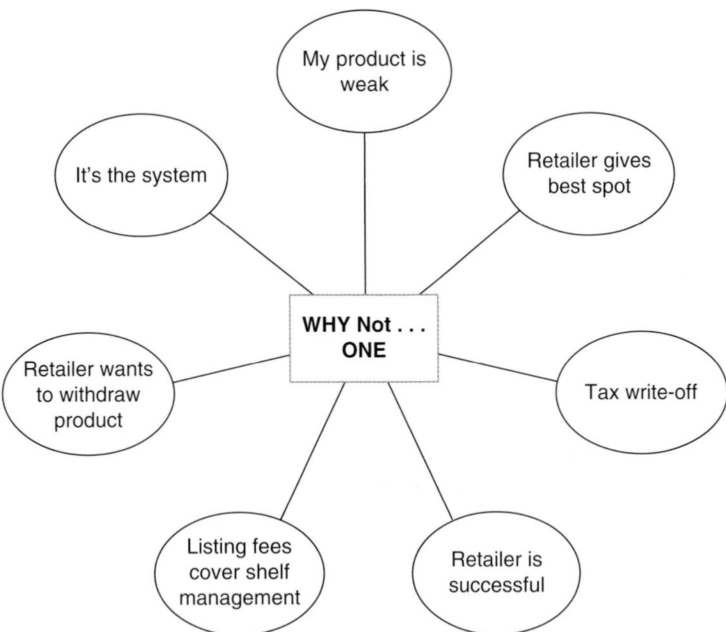

Figure 11.1 Retailers' new relationships: Why not number one?

- The listing fee covers stocking, removal, and restocking of my products, saving me personnel and money

And there are probably a few more explanations or excuses, which I have not listed and which you may want to add based on your own personal experience. I shall not discuss this list and its impact on the present-day relationships between food companies and retailers any longer other than to make two points: First, it appears that retailers are really in the business of real estate, and, second, more importantly for our discussion here, this clearly shows that a change of relationship toward partnership is dearly needed. This is exactly one of the three elements that compose the new business model 2.0: retailers become partners. I do realize that it's not an easy feat at all and takes much ungrateful legwork to start the rough journey toward real and equitable partnership between food companies and retailers.

Figure 11.1 illustrates the most prominent excuses why it is rather impossible to work with retailers without having to lay out cash for having one's products on the retailers' shelves.

11.1.3 Model 2.0: It's either all or nothing

I also strongly suggest that a let's-be-partners without the two accompanying elements of an R&D organization that drives the company and a wholehearted embracing of open innovation and innovation partnerships with appropriate external know-how providers won't work. It's got to be the three elements together, if not just to show the new retailer partner that strong and efficient change has happened in the food company. Let me discuss in some detail the consequences as well as possible flaws of and alternatives to business model 2.0 and what it might mean to the organization at large and especially the R&D group.

The two major consequences would certainly be the redefined relationship between food manufacturer and retailer as well as a totally redefined role for the R&D organization in the food company. The former might be heavily opposed by the marketing and sales organizations in the food company and the latter by the manufacturing and operations organization. Why would that be? Simply because of loss of power and weakening of old, established company "kingdoms" for some. The major arguments are:

- The proposed change has no merit.
- The proposed change adds no value.
- What is proposed is too radical and too risky; let's continue the old ways.

I kept the arguments on purpose on a generic level because you will always hear these same arguments—or similar ones—to fence off proposals for change. Let's, however, analyze these arguments in more detail.

11.1.4 We don't want to change anything; all is just perfect or is it not?

The first one, the proposed change has no merit is not so easy to debunk because merit is often in the eye of the beholder and even financial arguments are often dismissed over the hope to retain a cozy and well-established situation. It's so much easier to keep everything as it is than to try out something different. If there were a way through close partnerships and collaboration with retailers to substantially reduce or even totally eliminate listing fees, this would be a dramatic financial benefit for food manufacturers. I do realize that retailers apply similar arguments to avoid change, especially when it comes to retaining a financially profitable situation; why should they give this up? Well, maybe because they could even achieve bigger margins by collaborating closely with food companies and thereby rendering new product development more efficient and less costly, thus eliminating the need for rather artificial real estate rental fees charged for shelf space in addition to other elements of overall profit. The real weak point of my argument is obviously the need to get both parties to the negotiating table and to accept the new approach described in business model 2.0.

This should become easier if the manufacturing side had more convincing arguments to offer, such as introducing a fresh approach to the relationship between manufacturer and retailer by changing from the old and encrusted endless finance-heavy debates with the food companies' marketing and sales to having R&D as the new speaking partner, a group that is much more competent when it comes to knowing their products, and especially knowing their benefits for consumers. Today, R&D communicates such product benefits to marketing and sales, who, in turn tell the retailers and communicate to the consumers. Let's cut out the "middle-men" here; they are not really needed other than to assist R&D to find the right language when speaking to retailers and consumers.

The really big argument repeated time and again against this driving role of R&D is telling everyone who wants to hear it, and even to those who don't want to hear it that R&D people are typically not capable of leading an organization such as a food company; leave this to the grown ups and continue playing in your discovery garden. I do admit that there are certainly weak people to be found in the R&D groups, weak in terms of not being able to lead, but then one finds such people in every organization, in every department, and at every level. This is not an argument against R&D's suggested leadership role but rather an argument brought forward by those who are afraid that they might lose their power and privileges. In the end, it's just a question of hiring and training the right people, which includes new approaches to education at all levels to prepare the science and technology as well as engineering students to fill leadership roles.

11.1.5 It's about time for R&D to jump into the driver's seat

The suggestion for representatives of R&D to become leaders of a company is nothing outlandish in other industries, and you can see many examples in pharmaceutical as well as chemical industries. In these industries, it is quite a normal and non-frightening situation, whereas the food industry still believes (or rather makes believe) that only business people can run the business. Why? Fear of loss of power and privileges to the detriment of well-being and well-advancing of the company. I say it's about time that this changes, and R&D and their best leaders take over important leadership positions in food companies. Again, the three elements of the proposed business model 2.0 are complementary, and I would predict that for this model to become successful all three elements have to be adopted and applied. I had discussed the pros and cons of two of these elements—retailer partnerships and R&D driver—so let me emphasize, again, the importance of using open innovation and innovation partnerships as the crucial and unifying element to bring the two others successfully together.

Why do I suggest open innovation and innovation partnerships to be the glue that holds everything together? Simply because the attitudes of openness and acceptance of the roles that external partners can play and especially that intimate product knowledge are at the basis of every successful business in the food industry. This means that open innovation, innovation partnerships, and retailer partnerships as well as R&D leadership are held together by these attitudes, without which the model is likely not to succeed. There may be quite a few more arguments against or for the suggested model 2.0; yet again, the major arguments against were briefly listed—no merit, no value, too risky—and you can add more if you would like to do so. However, I would strongly encourage you to find more arguments to support the suggested model and make it a strong proposition that you may have in internal company discussions when the future of the company, and especially the future of R&D in the company, is discussed and questioned.

Figure 11.2 depicts the major reasons why business model 2.0 might meet resistance

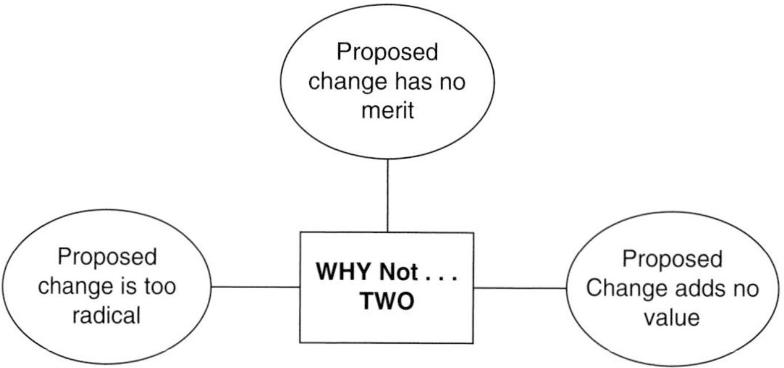

Figure 11.2 Typical arguments to reject business model 2.0: Why not number two?

11.1.6 What about business model 2.1? Too disruptive and outlandish?

As a reminder, the four elements of the suggested business model 2.1 are:

- The new food company sells know-how (in addition to products, eventually as the only product).
- Experts (scientists and engineers) are trained to become communicators.
- Experts work part-time off-site at retailer's space.
- Experts are additionally online to chat with consumers.

Again, are these really serious and practical proposals? Well, let's look into this in some detail and try to detect possible flaws and points that are potential value destroyers or a menace for the business overall. And again, much of the analysis is based on assumptions, which, in turn are based on personal experience and many years of observation. In addition, let's not forget that some of the elements in model 2.1 are based on the assumption that model 2.0 is at least partly implemented, especially when it comes to the new relationship with retailers.

The first big change that model 2.1 suggests is to say good-bye, gradually though, to selling products and to some degree services and saying hello to selling know-how, expert know-how that is. And such expert know-how is quite naturally linked to people, the experts, to make this a really credible proposition. It's pretty much living the "IBM Moment" in different ways so that a company such as Nestlé will not have to live the rude awakening of a "Nescafé Moment," just as the Kodak Company had its "Kodak Moment." The landing has to be prepared and should be a soft landing and not a hard and deadly one. This is the main driver behind the idea of the first element of business model 2.1, selling know-how instead of products. If this were to happen over night, one could immediately see the impossibility of this approach and that's most likely one of the major arguments against such change, especially brought forward by representatives of the manufacturing branch of a food company: "We sit on all these assets, the factories, what shall we do with those? It's just impossible!"

And yes, over night it is definitely impossible, but in a gradual fashion it is not only possible but necessary, mostly because consumers will increasingly turn away from industrial food products and will increasingly turn toward credible information and expert know-how linked to their food, their health, their nutrition, and well-being. The good news is that it will take a few years if not longer; but come it will and every food company, small or large, will have to get ready for such dramatic change so that a soft landing can be achieved.

11.1.7 So, what's bad about model 2.1?

Let me list possible flaws of a gradual transition away from being a pure product manufacturer toward becoming a real service and especially know-how provider. First of all, there is still a long list of manufacturing sites, which, in traditional ways occupy production lines, handled by (some) employees as well as robots that cannot just magically be transformed into thin air. This is a reality. On the other hand, if the predictions are right and consumers will increasingly shy away from industrially manufactured food, these production lines may become obsolete anyway. And factories, the bigger they are, the longer they live, as can be seen in the steel industry when looking at the present-day steel mills that date back almost a century, as in the case of the former Kaiser Steel, now California Steel Industries. Kaiser was founded during World War II and after a bankruptcy in the 1980s, parts of the machinery was sold to Chinese steel manufacturers and a good portion of the machinery is still in use to this day in Fontana, California.

Large equipment is difficult to do away with easily and has a tendency to stay alive for a long time to come.

Admittedly, equipment in the food industry is smaller than the large furnaces and steel rollers in the steel industry, but they are large enough to wanting to hold on to them. After all, it's really nice to watch a wonderfully functioning food production line and feel it's nostalgic aura. However, it might be good to prepare a gradual exit so that there is the least amount of financial disturbance and personal and personnel hardship to be expected. And yes, there almost certainly will remain a need for industrial food products or semi-finished food components that will have to be manufactured by someone; my prediction is that the real added value is not to be found in the traditional ways of selling industrial food products but selling real expert know-how around food and especially health- and wellness-related aspects.

11.1.8 We better start the gradual transition today

I don't take the pathway toward future obsolescence of large food manufacturing sites lightly because such assets, although largely written off in most food companies, especially the big ones, still have a nominal value that may be in the order of several \$10 to \$100 million per medium-size plant, so not really negligible. And there is much debate that the manufacturing industry is of great importance to the well-being of local communities because it supplies jobs. Well, I let you judge how much this statement holds still true today and how importantly a highly automated food production line contributes to the well-being of your local community job- or tax-wise or otherwise. The writing is on the wall that all of us need to learn new skills that are most likely outside the traditional manufacturing skills and that prepare all of us to fully embrace the present and even more future knowledge-transfer society.

I admit, I still have not solved the problem of what to do with the traditional food factory, and how much it will cost to discontinue the manufacture of food products and transition to this new reality of selling know-how rather than products. It will cost money; on the other hand, most of the assets are written off so the real costs will relate to people. I strongly suggest, however to fully embrace the suggested concept, and begin the smooth journey of gradual transition soon! And R&D, as suggested in model 2.0, not only can help in this, it will also be the new value proposition to consumers. This brings me to the next crucial element suggested in business model 2.1: experts are trained to become real communicators, those who can formulate and sell the real value proposition, know-how.

11.1.9 It's all about people

What are the flaws and weak points of this proposed element? Well, one can quickly see it: it's all people related, and people are more difficult to steer and control and remold than machines and robots. Yes, it's a trivial statement, but on the other hand, it totally describes what could go wrong: it's the people! This is going to be a real challenge and therefore makes the whole project of transition so much more interesting and potentially rewarding because the right people, having gone through the right education and training, can make the important difference to competition and potentially bring in greater benefits than selling products in the old-fashioned and traditional ways. I do admit that this is a critical and shaky element, and if people don't follow this approach wholeheartedly, well educated and even better trained, this entire experiment may just go south. It is therefore really important to organize the most appropriate educational and training programs so that the experts not only learn all about their expertise but have a strong desire to communicate and sell this expertise: selling of knowledge will become the new mantra for scientists and engineers.

And such knowledge is a living and developing entity and has to be complemented and supported by real practical bench work as to always keep the science and technology know-how at its highest level. Therefore, a good mix between practical ongoing persistent search and discovery, paired with the strong desire to communicate and sell is paramount. If the company gets it wrong, the model will not work or will perform rather badly. That is the real danger and will make or break this approach. It's all about having the right people, not only well educated technically but also being real champs when it comes to communication.

The centerpiece of the suggested business model 2.1 is of course the new orientation of the new food company toward increasingly selling know-how and gradually reducing the importance of simply selling industrial food products to consumers. The real potential flaw of this approach lies in the fact that it's totally unchartered territory for today's food companies; it is an unknown abyss! I have mentioned time and again in this book, the food industry is traditional, conservative, and resistant to change. So, how on earth would they ever contemplate, let alone accept, such a dramatic change like doing away with products and going toward a business model that has communicating and selling expert know-how at its core? Some or most of you may have the answer right away: never. But then, hold the horses, not so fast. Yes, I agree it looks and sounds impossible; however, if external events, such as increasing skepticism of consumers toward industrially manufactured food becomes more mainstream, then such forces can quickly drive a company to accept even dramatic changes. This is the purpose of proposing such a change: be prepared and ready should the need arise by embracing and gradually introducing such a business model.

11.1.10 Selling the intangible: The new mantra

Another important weakness of the proposed model 2.1 and the element of selling know-how rather than products lies in the uncertainty whether consumers are willing to pay for such a service. In times when services in other industries, from illegal music streaming, to illegal video downloads, are under massive pressure, it might be difficult to justify the idea that consumers would be willing to pay for food, nutrition, and health-related know-how. However, there are more or less successful examples in the food industry already in place such as Weight Watchers® or Jenny Craig®. Both approaches rely on the combination of specialized prepared products and expert advice that is given to consumers and for which consumers pay, either through the products, separately through a membership fee, or both. The case of Jenny Craig® is an especially interesting one because the company was temporarily owned by the Nestlé company and, after a few years of futile attempts to "nestlé-ize" Jenny Craig®, it was sold again. The model simply didn't fit into the mainstream approach of today's food companies, which is selling industrial food products. I shall discuss this case in more detail.

The two other suggested elements of business model 2.1 are straightforwardly complementary and a simple consequence of the basic elements. Experts who want to be communicators must go out to meet consumers and converse and interact with these. They can do this at many places such as food fares, food-related conventions and shows, or in schools and universities. Most of all, and based on a remodeled relationship with retailers as proposed in business model 2.0, they should be at the retailers' locations. This brick-and-mortar environment has to become the new, natural habitat of food experts. This should become the new and vibrant meeting place for experts and consumers, the daily mingling with consumers, and as said before, in combination with the regular bench-work so that the expert know-how is always relevant and up to date. And yes, they have to put time aside to be present in chat rooms and be ready to answer consumers' question patiently and in expert ways. This is probably the single biggest change that the R&D organization of the food company has to undergo: experts are no longer exclusively lab-bound, but they share times inside as well as outside the organization.

Figure 11.3 illustrates what is potentially considered bad about business model 2.1

11.2 THE NEW PEOPLE: WHAT DOES IT MEAN?

In the preceding chapter, I discussed the new scientists and engineers—a new type of people and suggested that such a new type does not simply appear out of nowhere but changes in education and training have to be at the base. Consequently, every educational level has to be part of this change and should start at a young age. Communication being the proposed new mantra, it has to be taught and acquired at every level and every age. By looking at the driver in front of me while waiting at the red light I can clearly see that she (or he for that matter) is

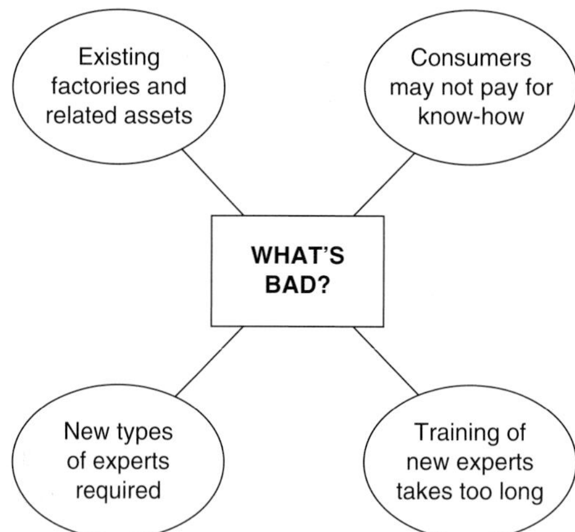

Figure 11.3 What's bad about business model 2.1: Why not number three?

communicating by texting some probably really important message to someone on the receiving end, like "I'll be home in 10, what's for dinner?" People seem to communicate all the time and yet also seem to have lost all talent for exchange of words other than in short text messages on their smartphones. This might sound like a rant against texting—and it should be when it's done in the car! On the other hand, many of us feel this strong urge to communicate just about every detail of their lives—unfiltered and live. But do they talk more, do they exchange ideas, do they listen and debate? My personal take on this is simply: no. We have to get this back: the desire to communicate verbally and directly in person!

11.2.1 Really new people with a new level of education are needed

This is the real flaw and weak point of suggesting a new type of direct communication, which is linked to the sheer existence of great communicators. They don't grow on trees and they don't arrive over night. Companies, if not the society at large have to push this need for great communicators into the curricula of schools at every level. Once this is achieved, at least in part, we can hope to get the "new people" into the pipeline and ready to go out and do their job, as would be needed in business model 2.1 mode. However, this is the most difficult part of the entire approach: achieve the necessary level of understanding and positive reaction of the world of education that the "new type" is not just a doer but both, doer and communicator. This is not an easy feat to achieve and depends on two assumptions: the educational structures and curricula are adapted accordingly and those who follow such curricula are willing to accept this new, added feature of becoming top class communicators.

However, it does not end there: companies who hire such dual expertise employees—technical as well as communication—must provide continued training on the job and most and foremost continued exposure to consumers outside the four walls of the company in the field.

This means that companies need to realize that such field jobs are not "lost time" in the daily work of the employees but are extremely fruitful and rewarding for the individual expert but more so for the company at large. One possible scenario of making this a desirable activity for the individual is to include a projected number of external missions in supermarkets, number of interactions with consumers, or frequency and duration of online chats with consumers in the target setting and annual objectives of the employee. Making it something measurable and auditable is a sure way of rendering this new approach as the new reality at work.

11.2.2 And there has to be more

Yet another potential flaw or weak point in finding, educating, and training this new type of talent required to take on the new challenges is the need to not only adjust educational curricula accordingly but also potentially introduce totally new and seemingly outlandish topics such as acting, drama, comedy, music, singing, or other apparently unrelated ones. Why do I mention these areas? Well, simply because they all have to do with direct verbal and "physical" communication. From personal experience I can add that I have learned most of my own communication skills quite a few years ago when I was the bass player of a small rock and folk band. The three other members of the band had let the honor to speak in between songs to me, and so I had to always make contact with the public and not only talk about the songs and the band but had also learned to crack jokes, tell short funny stories, and in general, entertain the audience in addition to the songs that we played. By the way, here's a good bass player joke:

> *The conductor stops the orchestra during rehearsal and says: Hey! You there in the bass section! You need to tune up! The bassists reply in unison: Is it that time of year already? [Anonymous]*

Joke aside, the point here is that to become a great and motivated communicator one has to acquire skills that are at first sight unrelated to the skill set that one already has. Just taking a course in communication (if such a thing actually exists) is by far not enough. One has to go out and really acquire these or possibly other appropriate skills. More importantly, curricula at universities and human resources departments in companies have to realize such needs and have to agree to make learning such new skills part of their offerings. What people learn today is boring and repeatedly results in the same type of boring employees, who had proudly accomplished their second or third similar sounding management course. Don't get me wrong: it's not profoundly wrong for a company to pay more than $50,000 for a six-week management course for their "high potentials," other than it's an awful lot of money just to make the employee learn and apply the company's "business speak." This is totally trivial and I have met scientists and engineers who were proudly applying words such as *granularity*, *revenue*, *profit margins*, *P&L statement*, *balance sheet*, *quantitative consumer research*, and the like. Again, there is an urgent need for R&D people to understand and live the business side of the company in much more intimate and shared ways than is the case today. But it is equally, if not more important, that scientists and engineers can properly communicate what is at the core of their work: expert knowledge and the excitement of discovery. This is what is really missing today and what is *the* crucial element of the suggested business model 2.1: experts are trained to become communicators and go out and interact to and speak with consumers.

11.2.3 Hiring by committee

And there is more and it happens strictly at company level: hiring of new talent. In the past—"the bad old days"—people were hired in all possible and impossible ways: they knew someone in the company, the manager who wanted to hire interviewed and liked a particular candidate, the group in which the candidate was going to work in got to know him or her during informal short impromptu meetings, and so on. What I am trying to get at is the fact that at the end the decision was taken by the manager in charge, and he or she hired the one person they liked best and who they believed was best fit for the job.

Typically, at least in larger corporations, this is not the case any longer. Hiring-by-committee (HBC) has become the new etiquette. With all due respect, this is really abdicating responsibility of the individual manager. Everyone hides behind the curtain of the committee and the whole process becomes extremely impersonal and anonymous. It's about time to go back to the bad old times and make hiring a personal, emphatic, and daring endeavor again, in which managers take personal responsibility and can be held responsible and accountable for both success and failure in the hiring process. The more recently encrusted attitudes by management committees when it comes to hiring talent is probably one of the bigger obstacles to successfully changing the work force to become communicators in addition to the traditional trades that the experts represent.

The way out is really more personal responsibility of managers when hiring and fostering an attitude of more gut and less committee and especially moving away from hiring talent that strives to please the boss toward hiring talent that is relentlessly pursuing the idea of serving the cause; that cause being one that includes the new tasks proposed in the business model 2.1, in addition to supporting the elements outlined in the proposed business model 2.0. This is not an easy feat and requires the recognition of management that changing attitudes goes hand in hand with changing and continuously improving direction and content of the company's DNA.

11.3 SOME CASE STUDIES: PERSONAL VIEWS

In this section I shall analyze and discuss a few cases of types of businesses that were pretty close in their approach to what was suggested in the new business models 2.0 and 2.1 and which have been incorporated—I would call it *devoured*—into a large, traditionally operating food company. I use examples that I know best because it happened almost "next door" during my time with the Nestlé Company. Let me begin with some fun or at least with a family of brands that represents fun and playfulness as the major brand values. I refer to the Willy Wonka® brand, which was crafted after Roald Dahl's classic children's novel *Charlie and the Chocolate Factory*. The Willy Wonka brand was eventually launched in 1971 and the Chicago-based Breaker Confections company, owned by the Sunmark Co., which in turn was a subsidiary of Quaker Oats®, but Sunmark Co. called the shots and successfully developed products under brand names such as Nerds, later Nerds Rope, Laffy Taffy, Everlasting Gobstopper, Sweetarts, and other products originally mentioned in Dahl's novel.

11.3.1 *Charlie and the Chocolate Factory*

Sunmark Co. was eventually acquired by Nestlé in 1988 and, as usual for the Nestlé Company, it attempted to imprint Nestlé on Sunmark and the Willy Wonky brands from the beginning. When I first met with representatives of Sunmark a few years later, I was exposed to the most important mantra for the Willy Wonka® brands when it came to new product development. When the management team tested new product concepts, and they generally liked a specific variant, it was clear that this variant was not going to be pursued because it was the opinion and taste of adults and most likely the exact opposite of what kids would like. In turn, concepts were successfully pursued that were generally really disliked by the management team. Coming from the traditional new product development approach of choosing product concepts that were liked by a group of decision makers, I was rather surprised by this approach and would not dare to argue with the success that Sunmark and Willy Wonka experienced in those years.

As mentioned, the large mother company little by little imprinted its DNA and the traditional ways of new product development. You can probably understand that during the years that followed, the Willy Wonka® brand suffered and product successes were rare and new product development was all over the place. It took a few years until Nestlé finally realized that letting the reins loose and follow a rather unorthodox approach to new product development was the better and more successful approach, and the brand has become rather successful again.

I use this example simply to illustrate that too much control is often counterproductive and achieves the opposite of what is hoped for. Letting the R&D group more freedom in turn means that the scientists, engineers, and product developers have to play a different, more independent role, which typically includes strongly improved communication skills to better explain their work, both to management, which decides on budgets and supports, and also ultimately to consumers.

11.3.2 It's all about talking to clients

Although this example has more or less experienced a happy end—rather a happy intermediate—my next example does not end so happily. I refer here to the case of the Jenny Craig Company, which by its original definition perfectly fits into elements of especially the suggested business model 2.1 (i.e., scientists being solid communicators of expert knowledge to consumers, the clients of Jenny Craig). The company was founded in Australia as a small family enterprise in 1983. There was and is actually a Mrs. Jenny Craig, who gave her name to this enterprise. From the onset, the business model was selling expert advice in the area of weight management, weight loss only being one part of the endeavor. Two years later in 1985, the company expanded abroad and settled in Carlsbad, California. I had the chance to visit them a few times and was rather impressed by the attention to their clients in terms of discussing their needs and nutritional requirements to lose weight and even more importantly manage weight loss and avoid the often experienced situation of regaining lost weight. There are other companies who have similar approaches, however, as mentioned, I have personal knowledge of how Jenny Craig evolved under the ownership of the Nestlé company, which I use for the purpose of this analysis. The case is also dear to my heart because it is the model that comes closest to the suggested business model 2.1 and that proposes exactly this: sell expert know-how to consumers or, as in the case of Jenny Craig, clients.

11.3.3 Observe and learn; don't impose and remodel

Between 2006 and 2013, the Nestlé Company owned Jenny Craig, Inc., and old habits die hard, the "nestlé-ization" began rather early on. What do I mean by that? Well, those of the readers who have lived through similar "-izations" probably know exactly what I mean: control and imposing mother-company rules and ways of doing business, by all means and at any rate. Even, if that could potentially destroy the corporation that was purchased dearly; in the case of Jenny Craig, it was speculated that the transaction was about $600 million worth. My visits and meetings with Jenny Craig in Carlsbad came rather late in the relationship in the early 2010s and it was already apparent by then that the preliminaries for a divorce were in the works. The overly zealous control and a deep misunderstanding of the business model, a model that did not typically fit in the Nestlé model of doing business, were the major drivers towards this split, probably paired by increasing mistrust and disappointment with a situation that had turned from bad to worse.

My personal analysis of the situation is not complex: the particular business model that Jenny Craig had developed, namely directly talking to their clients in personal meetings and selling mainly expert know-how and support, did simply not fit into the traditional business model of the Nestlé company, which is based on developing, manufacturing, and selling prepared food and beverages. The "marriage" could, however have worked had Nestlé had the guts to admit that they were not in a position to impose their ways but rather observe and learn new ways of doing business with potential long-term positive impact and the starting point to reshape their conservative and old business model.

I should add, though that such separations are not only happening on a single level but many things come together, which are often linked to people, personalities, and egos, and that was certainly the case here too.

11.3.4 *Citius, altius, fortius*

Let me discuss a third example here, which I again know from personal experience and many years of observation of what had happened: the case of PowerBar, Inc. Just a short reminder, I am using these cases in my analysis because they all have business models that are based on similar assumptions as suggested in my business models 2.0 and 2.1. PowerBar was created by a long-distance runner who was not happy with the energy-delivering and small and convenient products available around 1983. The founder and creator of PowerBar, Brian Maxwell, was an Olympic athlete and was especially dependent on products with high energy density, convenient to eat while running. Allow me a personal note here: my late father in-law was a mountain climber who had done many summits in the Swiss Alps during his "hay days" of climbing. This was quite understandably long before energy bars became fashionable, but he, like all the other mountain climbers had their personal favorite energy bar: a tablet of milk chocolate—100 grams of milk chocolate represent approximately 600 Kcal energy. A good portion of around 30 percent is fat, highest in energy density; most is sugar and some is proteins. In the light of today's nutritional know-how and especially detailed know-how when it comes to the needs of athletes while competing, I do agree that a tablet of chocolate might appear as a rather crude approach, but nobody can argue with it being light, rich in fast available energy and well tasting. The latter, with all due respect, cannot really be said of most of the energy bars: these are moderately well tasting (euphemism!) and especially the high protein bars have a tendency to pull your teeth's fillings.

The principle business model of PowerBar was to directly work with athletes and discuss their personal, often individual nutritional requirements both in the period running up to a competition and especially during competition. Although there are certainly individual differences when it comes to specific energy needs, it was quickly realized that there is an overarching similarity, which allowed them to create generic energy bars that were all particularly rich in proteins. This was probably one of the reasons why Nestlé acquired PowerBar in 2000, and same playbook here, pretty quickly began to "nestlé-ize" the newly acquired child, PowerBar. Chances were that PowerBar could become a good fit for the company, as in those years, the company had a strong drive toward making important marks in areas such as adult nutrition and especially performance (i.e., sports) nutrition. The purchase price was, given the size of Nestlé, rather moderate and was estimated at just below \$400 million and at an annual revenue of approximately \$150 million this looked like quite a good deal, with great potential to grow into new geographies and beyond elite athletes to the average recreational athlete, even the energy hungry manager in dire need of a boost. So things looked promising in the beginning, and it is difficult to pin the outcome of separation after more than a dozen years down to one reason. Like in the other cases, the business model was selling expert knowledge in the format of a nutritionally designed energy bar. Another reason could have been the success of the energy drinks segment in the marketplace, especially the big success of Red Bull®.

11.3.5 Some reasons for the separation

My personal favorites why this adventure has not really worked out follow. First, you can only do so many iterations on energy products—there is only one Red Bull®—and PowerBar probably had too many product iterations. Secondly, although taste and texture are generally personal, it has to be said that most of the PowerBar products didn't really please the palate and were extremely difficult to chew and swallow. Thirdly, and again this is a personal observation, there were too many personnel changes, especially on management level. When one person in charge did not succeed fast enough to grow the brand he or she (typically he) was replaced quickly. This certainly did not help the cause.

After more than a dozen years of operating in the Nestlé family, PowerBar had allegedly increased the annual revenue by a measly 17 percent or so to \$175 million or approximately 1.3 percent per year. Inflation was greater during those years and so here is my last and fourth reason why the relationship was finally ended and why PowerBar was resold for approximately $400 million in 2014: management was asked to "go back to basics." It is a well-known fact that "going back to basics" in business means the declaration of defeat. So, shortly after this order of going back to basics, PowerBar was sold by Nestlé.

Overall, one could argue that not much was lost in financial terms; however, I would argue that a golden opportunity to real venture out toward a business model that works with and talks to consumers was dearly missed and the company has been "poisoned" to go into a similar area any time soon again.

There are probably many more examples that could be discussed in the context of missed opportunities to try out new and different business models, especially the ones suggested in the preceding chapter. And it would be about time for food companies to try out something new because the old ways, the models that are close to 1,000 years old, become really outdated and increasingly disliked by consumers.

Before ending this chapter and the arguments for and against new, disruptive business models in food companies that in turn would have a dramatic, almost revolutionary effect on every company's R&D organization, let me add one more thought, one more model that exists in other industries, at least to some degree, and that could be the new way that an R&D organization, especially in the food industry, ought to go.

11.4 BUSINESS MODEL 3.0 FOR R&D

As if there weren't enough disruptive proposals yet and described in some detail, here I offer a third suggestion, business model 3.0. I only bring it into the discussion toward the end of this chapter, which is all about critically discussing the options and weighing their merits as well as potential weak points or flaws. The reason for this is simple: the model already exists and has some history on its back already. I speak about the well-known Palo Alto Research Center (PARC) that, in its heydays was the internal R&D organization of the Xerox Company. It was internal until the day when cost factors forced the PARC to become more independent, at least in part. This meant that one part of the projects that PARC dealt with were no longer exclusively carried out for Xerox but for third parties in noncompeting areas. This sounds easy but, at least in the early days of the separation must have been a logistical and especially IP nightmare.

Xerox PARC was founded in 1970 and was world renown for its advanced research program and especially its groundbreaking results. Many IT and computer-related breakthrough discoveries were made by PARC and its dedicated research teams. Many well-known personalities in the world of IT such as Robert Metcalfe, founder of 3Com and Ethernet, and John Warnock, founder of Adobe were researchers at PARC. Others such as Steve Jobs of Apple and Bill Gates of Microsoft were part-time interns. And there are many more examples of highly creative and entrepreneurial people who came out of PARC to create their own companies and in addition gave PARC an extremely high standing in the world of IT-related R&D. This apparent openness was of course great for the reputation but maybe less good for Xerox. The mother company had a long lasting focus—too long lasting perhaps—on photocopy machines that, with the exception of the laser printer, not too many other inventions that came out of PARC were used within. The ultimate great successes were for other mostly competing companies.

11.4.1 Change was in the air

It was therefore only a matter of time until Xerox recognized that their approach to PARC and the underlying business model had to be changed. The reason for that was simple: when less comes back compared to what is invested, decreasing returns require dramatic change. There are two possibilities: either the activity and necessary investments are discontinued or they are reoriented. In the case of PARC, this reorientation led to the decision that the research group became a subsidiary of Xerox back in 2002 and had to find clients to support and finance projects outside the sphere of interest and influence of Xerox. My personal experience with PARC when discussing potential research projects for my former company date back a few years and the observations are most likely not totally up to date. In those years, around 2008, PARC had a 50/50 split of projects requested and financed by Xerox and third parties, respectively. Because third-party projects by definition had to be outside Xerox' area of interest, it is clear that to be able to run other companies' projects in different areas, new disciplines and know-how had to be added to the research portfolio of PARC. One such competence area was water purification by mechanical means and is a great example of what PARC could and still can do for companies who need specific competencies, which they do not have within their own internal portfolio. I would call this mixed structure "brain for hire" or maybe better "competence for hire."

I received the following additional insight from a former member of PARC's research staff (personal communication, Nitin Parekh, September 2015):

> *In terms of numbers of researchers it had to downsize from an estimated 300 to an estimated 170 or so. Research staff is organized across 4 labs. The major driver for doing this came from some tough times that Xerox went through in those years leading up to the transformation of PARC in 2002, especially the feeling that PARC had much more than what it needed for its own good. Cost cutting was of course the underlying element for this hybridization of PARC. To this day, the project split is still estimated to be 50:50 internal (for Xerox) versus external, for third parties. The model is still considered being a successful one and has definitely avoided the dismantling or dissolution of PARC. However, living this model is not an easy feat and only time will tell whether the new PARC can be considered a success story. Properly run and managed, such a hybrid organization can be a very valid model for others.*

11.4.2 A short commercial

Let me quote how PARC sees itself today, and which briefly and clearly describes this shift from full integration to partial "freedom":

> *PARC, a Xerox company, is in the Business of Breakthroughs®. Practicing open innovation, we provide custom R&D services, technology, expertise, best practices, and intellectual property to Fortune 500 and Global 1000 companies, startups, and government agencies and partners.*

We create new business options, accelerate time to market, augment internal capabilities, and reduce risk for our clients.

Since its inception, PARC has pioneered many technology platforms – from the Ethernet and laser printing to the GUI and ubiquitous computing – and has enabled the creation of many industries. Incorporated as an independent, wholly owned subsidiary of Xerox in 2002, PARC today continues the physical, computer, and social sciences research that enables breakthroughs for our clients' businesses.

The real difficulty of the setup is the potentially logistical nightmare of separating competences and making sure that each company's IP rights are protected. It's easy to hang blue plastic sheets in factories when running a line for a competitor to fence off curious eyes; it's another feat to clearly separate competences applied and results achieved in a setup such as at PARC. However, this could be the future of R&D organizations in the food industry. R&D has to serve two masters: the food company that owns them, but they also have to find research projects outside, which are requested and funded by other companies, not necessarily competing ones but companies which might need similar competencies to solve their own, specific problems. And, like in the case of PARC, new competencies might have to be added over time, which evolve from the existing base and venture out to satisfy the new, enlarged requirements. This is my suggested business model 3.0 for the R&D organizations of food companies, or probably any other company that still operates in the fully internal mode. It requires a major change to the present-day approach, especially in times when open innovation and innovation partnerships in some food companies are still seen as endangering the full grasp of the already fairly meager value chain. However, before R&D organizations are completely downsized to a ridiculous residual size without teeth and claws or before the disappear completely and everything is outsourced to competency providers who have no or little know-how and understanding of the intrinsic needs of the food industry, it might be the better option to head for a hybrid solution such as perfectly demonstrated by PARC.

Figure 11.4 describes the transformations needed when applying suggested business model 3.0

11.4.3 Change or perish

I do realize that not everything that happens in this hybrid example is perfect, but in my opinion it's the right way to go and is also totally compatible with the two other business models 2.0 as well as 2.1. One thing has become clear to me in the many years I have worked in the food industry, most of these years in an R&D environment: everything changes all the time, nothing stands still and nothing is forever. You may, rightly so, argue that this is a trivial observation, and I do agree, it is. However, one of the strong sentiments that one finds in most R&D groups in a food company is denial. What I mean by that is simply the fact that although scientists, engineers, technologists, and every other kind of expert who works in R&D mentally accepts the need for change, yet they emotionally reject it and often outright deny it. This has much to do with wanting to preserve the coziness and apparent stability that can be found in R&D

FROM	TO
Projects exclusively for own company	Project portfolio mixed for own company and third parties
Expertise and IP exclusively internal	Expertise and IP shared with third parties
Absolute exclusivity	Shared results
Closed innovation	"Invert" open innovation → R&D becomes know-how provider for others

Figure 11.4 Business model 3.0: The required transformations.

organizations. It might come as a shock to researchers and developers that they might have to compete for projects and get their financial support from third parties and not from their well-known and comfortable own company. And these are probably two of the biggest flaws and drawbacks of the business model 3.0: aversion to change and fear of the unknown. This is of course not unique to this situation but is a universal problem. However, R&D should be the first ones to embrace and accept new and partly unchartered territories, especially when it comes to their own survival.

11.5 SUMMARY AND MAJOR LEARNING

This chapter was mostly about critically discussing the suggested new business models and testing the put forward hypotheses. Some practical cases that had similarities to the proposed business models were discussed; some of these cases reflected failures, whereas others ended more successfully. Finally, a third business model 3.0 carved after an existing model was introduced. This model 3.0 is a hybrid between two scenarios for R&D: fully internal versus fully outsourced and could be a successful path towards the future of R&D in the food industry as well as other industries, which struggle with the same dilemma how to support and finance R&D in the future.

The following describes the major talking points and learning from this chapter.

- The suggested business model 2.0 was critically analyzed and discussed, and its individual elements—retailers become partners, R&D drives the company, open innovation and partnerships are fully embraced—were tested individually.
- The first element—the retailer partnership—can easily be dismissed by many obvious arguments; most of them too obvious and too easy to use. It was suggested that most of

the arguments are mere excuses and it should not be too difficult to overcome them, provided there is a strong desire to do so.

- It was suggested that model 2.0 can only really work if all three elements are wholeheartedly accepted, although it can be assumed that a gradual transformation toward 2.0 can be undertaken successfully with retailer partnership being the first step.
- General arguments against the introduction of model 2.0 could typically be the following: the proposed change has no merit; it adds no value; or is too radical and too risky. All of these arguments were debunked in the following discussion with the main arguments of change aversion and run in and accepted relationships with retailers.
- It was discussed that the proposal of getting R&D into the driver's seat of the company is probably the one that causes most fears and automatic refusals by most other parts of the company, especially the operational divisions. It's mainly a power game with the desire to hold on to privileges that suppress any possible changes; much effort and conviction is needed to change this direction.
- It was mentioned that in other industries, such as pharmaceutical and chemical, an R&D drive is quite natural and accepted for a long time. Accepting an R&D leadership in the food industry would perfectly well go hand in hand with the necessity to open up toward the outside, especially when it comes to the relationships with retailers as well as external competence or solution providers as is the case in partnership mode.
- Business model 2.1 was also critically discussed and tested. The four basic elements of model 2.1 are selling know-how instead of products, experts become communicators, experts work off-site part time, and experts also work on line and communicate with consumers directly.
- It is fully understood that suggesting to going away from selling products toward selling know-how is not easy to accept for many or most in the industry. However, with a growing concern and suspicion by consumers toward industrially manufactured food products, it was suggested that it may be about time to begin a gradual transformation to avoid a future crash and prepare for a soft landing. The biggest flaw of this scenario is the fact that food companies typically own a large number of still valuable assets in the form of factories and manufacturing lines as well as warehouses, so a hard transition is virtually impossible; hence the proposal to do this gradually and begin soon, especially to learn the "tricks of the trade."
- To sell know-how, the experts will have to become not only those who ultimately sell but have to especially become top-class communicators. The biggest problem with this approach is again the time it takes to achieve this through university education, in-company training, and especially new hiring practices. The new product is an intangible one, and this is of course the other important flaw: will consumers accept such a product and will they pay for it? That is the real question and no definitive answer can be given today. The model has to be tried out. However, there is a strong reason to go in this direction, namely increasing criticism of industrial food products by the average consumer.
- It was discussed that model 2.1 largely relies on people and how (if?) they can be motivated to accept the necessary changes that come with the new approach. Will the experts

accept to work totally differently (i.e., to work part time gathering and polishing up their know-how base and on the other hand be ready to work off-site and online)?

- Consequently, it was suggested that a new type of people is required to make business model 2.1 really work and successful. This in itself is the weakest point of this model because people need to be formed, motivated, and properly used. Most of all, such people need to be or become available.
- Another weak point related to the new type of people lies in today's approach to hiring new talent. Instead of managers taking responsibility when it comes to hiring, such responsibility is diluted and hiring by committee has become popular. This is counterproductive to successfully introducing the business model 2.1.
- Some case studies illustrating similar approaches found in the two suggested new business models 2.0 and 2.1 were discussed. It could be demonstrated that some of the cases discussed were rather successful while already using some elements of the suggested models and others were not. This had much to do how much freedom of action such companies had as members of a large food company, in this case the Nestlé Company.
- At this point a third possible model how to operate R&D in a company was introduced. Business model 3.0 is based on the example of the Xerox Palo Alto Research Center (PARC). PARC had transformed from a research center fully dedicated to its mother company Xerox to a unit that has a mixed portfolio of projects, partly for Xerox and partly for external third parties.
- The major drawback, if there was any, is the difficulty to clearly separate know-how applied for the mother company and know-how and resulting IP for external clients.
- Finally, it was mentioned *that the* major drawback for any type of suggested (or not suggested) new work models either for the company at large or for its R&D organization is caused by a general aversion to change.

REFERENCE

"PARC today," retrieved from http://www.parc.com/about/.

12 Summary, conclusions, learning, and outlook

Even for being stupid you need brains.

Karl Farkas

12.1 THE TYPICAL R&D ORGANIZATION IN THE FOOD INDUSTRY

I am not insinuating in the slightest that you are stupid; it's rather directed toward myself because it sets the stage for this last chapter in different ways. It probably requires some naivety paired with the uncompromising belief that things can be changed, even if they seem to exist forever. I prefer to call this mix *impatient stupidity*. Impatience together with stupidity or rather asking the so-called stupid questions is what drives this entire book. At the beginning of this book, I recounted a story of R&D in the food industry, what it was, what it is, and finally what it could, or rather should be. It's a fairly long story but should probably be so much longer when looking at the future. But let me begin at the beginning. Let me begin to summarize some of the personal observations and experiences that I had described at the start.

R&D is considered to be the underdog, and management does not let an occasion go by to denigrate the R&D organization as a whole and especially its people. This is not an invention by a frustrated and ill-understood former R&D guy, but it is an observation personally experienced time and again. Business managers in the food industry seem to have a favorite pastime: bully R&D and blame everything on the expensive, slow, and nonresponsive R&D organization in "their" company. If they are already paid too much, at least they should have the good sense to serve as scapegoats. I have observed many such denigration sessions and, after some time, one realizes that the ritual is always the same and is mainly based on the assumption that R&D people have no sense of business, they just spend money and don't come back with usable and useful results or at least not fast enough. Having done some psychology at university, admittedly many years ago, I would describe such behavior being largely driven by an inferiority

Food Industry R&D: A New Approach, First Edition. Helmut Traitler, Birgit Coleman and Adam Burbidge.

complex, and quite a big one. Business managers would never admit to such mundane feelings, but it was always quite clear to me; they needed a defense mechanism to explain that marketing, sales, and other business functions were so special, so impossible to learn for anyone other than a business talent (apparently this is in the genes) and R&D people would never understand the professions of business.

12.1.1 You are too old for marketing

The funniest excuse or attempt for an explanation why R&D people never could become marketing experts was: they are too old! One has to become a marketer at a young age, almost right out of kindergarten or not much later. Just imagine, what a lame excuse! And to my shame I must admit that I had actually believed it and used it myself when managing R&D groups, trying to explain curious and ambitious R&D experts that marketing was out of their reach. Shame on me for having believed and repeated such a nonsense.

My first personal encounter with corporate R&D, or rather, R&D work commissioned by a corporation and performed by a university department happened many years ago when I was a rather young research assistant at university level. What I have discovered then and what still stuck with me to this day is that those who have power over budgets, thumbs up or thumbs down in a kind of Julius Cesar fashion, are typically the business people of the company and R&D has to deliver, based on the amount of money that is given to them. It was said that almost by definition, the money spent on R&D in the eyes of the business is always too much; it appears to be a natural law built into the business. Again, I am not complaining, although I seem to do so; I am merely recounting observations, experience, and simple facts. However, it is a fact, at least in food companies, that their R&D organizations are kept on a short leash, for many reasons; some of them are justified and understandable, but the majority simply for reasons of fear that the smarter R&D people could find out that business is not enigmatic and not difficult to learn. It is a fact that becoming a scientist or engineer takes much longer than becoming a business expert such as marketing, sales, finance, or generalist leader.

Strangely enough and despite all of this, R&D organizations still exist, and probably stronger and with more experts working in R&D than ever. Yes, there is the constant drumbeat of "we have to reduce the number of employees," and it's an almost boring iteration. It seems to go on since forever, or even longer.

12.1.2 How it all started

A bit of history was discussed, especially how R&D organizations in the food industry were created and grew out of fear of resource scarcity, in particular food and other environmental resources. The "limit of growth" after a lengthy period of cheap and available oil turned into a period of fright and the infamous oil shocks of the early 1970s chimed in a new period of a more careful management of finite resources. During the days of abundance in the early and mid-1960s, oil was used as a raw material for food, proteins to be precise, and this was also the time when large food companies invested in food- and beverage-related R&D. Although it was quickly clear that such proteins from oil did not have much of a future, R&D grew and thrived and had to diversify its activities into the major food-related areas, often simply related to the major food ingredients and their functional counterparts.

This was the period during which research areas evolved from a singular approach to one large research focus toward multiple, smaller areas. This approach toward a large number of research areas led to a diversification of required competencies, and at the same time, leading to an almost inflationary expansion of size of R&D groups as well as number of projects. The risk management changed from one or just a few large projects supported by almost all resources toward spreading the risk and the resources across many smaller projects, thereby hoping that not all would be lost in case some of the projects wouldn't deliver. This led then to a belief of management that R&D was not only not efficient and slow to deliver but also considered to be eating up too many resources of the company. An endless series of cost-cutting and resource-reducing exercises began and are still ongoing to this day. Reorganization became the new mantra with the dire consequences that scientists and engineers in the various R&D organizations always seem to focus more on personal survival than on delivery of results.

Another important question discussed was: why does the food industry need R&D in the first place? Several answers were given and discussed. The most obvious answer is simply because they can. This is certainly only a superficial answer and several answers hide beneath such as: there is a tax write-off, the consumers indirectly or directly request it, and food companies were originally built on technical discoveries and breakthrough technologies. The consumers not only want their preferred food company to run a R&D group for strictly functional and food-related reasons, but in doing so, the standing of the company is heightened in the eyes of the consumers. Another answer is that the generation of proprietary intellectual property (PIP) represents a great intangible value, especially in the books of the financial analysts and consequently of investors. The more recent mantra of most food companies is the one that puts the consumer on a pedestal and defining the consumer to be the king, not having fully grasped that the ultimate consumer who makes decisions may be a queen.

12.1.3 Why R&D?

Here's an overview of answers to the question why the food industry needs R&D:

- Because it can.
- Because it always had R&D.
- Because it's origin was based on a product idea developed by an R&D like group or individual.
- PIP generated by R&D adds value.
- It has possible tax advantages.
- There are regulatory requirements.
- The company is serious about the quality of its products.
- Ultimately: the consumers demand solidly researched-and-developed products and services.

12.1.4 Everything's a project

The importance of projects for companies and especially food companies was discussed. It was said many times already elsewhere that all R&D work is strictly project based. But what is a project? Everyone seems to have an answer; however, the answers may differ and there may be different definitions to what *project* means. Here a list of key words that spring to mind when trying to define project:

- Collaborative or collaboration
- Involves research and design
- Carefully planned
- Aiming to achieve a particular goal
- Projects are temporary
- Projects go across organizations
- They need to accomplish particular tasks
- They happen under time constraints

There was a time, not so many years ago, when project was defined as a teamwork activity. Although there is definitely merit to the notion of teamwork, simply because one person cannot have competencies in all disciplines required for a project, it was probably hopelessly overrated, which led to much confusion as to who did what and was responsible for which part of the success story. It also led to another phenomenon, namely using projects as objects to hide behind and ultimately to excuse one's own weak points and potentially blaming other members of the team.

12.1.5 And here came the strategic business units

This was also the time when another phenomenon came over the food industry, pretty much like the flood over Noah's ark: strategic business units (SBUs). It all started in the early 1990s with one of the more aggressive consultancy companies having been more successful than others in selling the idea of combining all product group–relevant activities under one strategic leadership group. Although the basic idea was certainly a good one, it still begs the question as to how efficient this setup ultimately was and ever could be given other existing structures and complexities in most food companies. Reality was that the heads of SBUs never could really decide anything of importance. For instance, in the case of confectionery and for the case of the Nestlé Company, only approximately 30 percent of all confectionery products fall under the definition of being global and strategic, and only those would be managed by SBUs. Moreover, *managed* only meant the power of the head of the respective SBU to say *no* to a new layout of the brand layout on the package or say *no* to introducing a new ingredient, which may have altered the taste too much. The ultimate power to say *yes* was and is entirely with the product group in the respective market, and yes, after consultation with the relevant SBU representatives.

There are many approaches to be found when it comes to projects and especially project management. Every new approach tries to fix the flaws of the previous ones but mainly makes everything more complex and complicated. Rocket ships, pipelines, innovation funnels, end-to-end project management, handover of projects, quick-and-dirty, and more are typical expressions found in the context of project work, especially in R&D environments. And there is more such as conceptualization, preliminary exploration, quick-and-clean, enough to make your head spin. The major learning of this is simple: terminology does nothing to alleviate confusion, and yet another new approach will not solve the doubt that management has in the efficiency of R&D.

12.1.6 Clever project management

Project management was another topic that was extensively discussed and the question was asked: manage or lead? Interestingly enough, it is common belief that someone who manages a project is of higher status than the one who leads a project. And in reality this is the case, albeit the fact that leadership in projects is so much more valuable than managing a project: leadership takes risk, management glosses over them. The central question, however, is: what is good project management and a list of elements was developed that attempts to give answers to this most important question. Project management is mart selection of a project, mainly based on credible content and realistic expectations, as well as the right group of experts to work on the project—internal as well as external resources.

Most importantly, anyone who works on projects should ideally select the right project and make sure to deliver the promised results on budget and on time!

12.1.7 The role of the SBUs and how it influenced R&D

Another important topic that was discussed in some detail was the fact that actually all projects are sponsored by someone, normally a business or a business unit, typically by marketing as the sponsoring agent. Oftentimes so-called "skunk" projects are carried out in some kind of obscurity, and bosses typically don't want to know about these, although they definitely want to stand in the limelight once such a project became a great success. Even such projects are indirectly sponsored by the sheer fact that those who work in such a project are paid by the company, at least in part to work on new ideas of their own, which are not necessarily directly sponsored by any type of business organization.

Some in-depth space was given to discuss the role of SBUs in a food company and how they interact with or influence R&D at large. From personal experience I could add that the process of sorting out the exact respective roles of each partner was not a trivial one and much time was spent to hash it out, with the end result that some wordy type of contract was ultimately to be found on paper but reality looked a bit different. Although a difficult exercise, it was certainly a worthwhile one because it cleared the playing field, at least to a large degree. However, it also left a rather large amount of bias and overlap, which ultimately did not really help to always sort out disputes or unclear situations.

The following list of typical interactive tasks between a SBU and an R&D group was established and discussed:

- SBUs gather consumer understanding and insight; R&D defines and perfects relevant technical skills.
- SBUs develop business relevant strategies; R&D assures the corresponding expert resources.

- SBUs write briefs based on insight and strategy, ideally in collaboration with R&D; R&D organizes and sets up appropriate projects and project teams.
- SBUs send briefs to R&D and request their execution; R&D delivers results according to plan.

12.1.8 The rituals: Consumer research, business plans, and the project definition

Role and importance of market as well as consumer research were discussed in much detail. Consumer research is widely used in the industry; however, I believe that it is often taken as an excuse and almost alibi to gloss over the weakness of marketing plans. Business plans and the resulting steps to take are of course at the origin of all project work and should therefore be well communicated, especially to the R&D organization so that relevant projects can be defined and carried out in the best possible and most efficient ways. Even the best of projects always seem to have some kind of hiccup and typically take longer, if not much longer than initially planned. Some projects never seem to die and seem to be resurrected frequently and under new names. Project review meetings are held for the purpose to check the proper advancement of projects but often are just another ritual that many just want to get on with. Although project reviews would also serve to terminate unsuccessful projects, I have personally only rarely observed this.

As a reminder, here the list of ground rules to apply to make project review meetings as efficient as possible:

- Be factual.
- Be short and concise.
- Always refer and compare to agreed-on milestones.
- Tell the truth, even if you would prefer not to.
- Never go over board with visual support.
- Don't be defensive.
- Make sure that you get a decision from management.

At the end, what really counts is a successful outcome of any project, especially with subsequent success in the marketplace. This requires a commonly agreed-on set of success metrics, which is not always easy to have. I have experienced too many excuses why projects, despite having hit all (or almost all) milestones, did ultimately not succeed in the marketplace, such as the consumer research was misleading, the resources were not available fast enough, the business changed time lines, and a few more. At the end, none of this is of any relevance other than it most often will give R&D bad press and a bad reputation.

12.1.9 A critical view of today's R&D organizations in the food industry

I also critically analyzed and discussed structures and people in a typical food industry R&D setup. Historically, R&D organizations in the food industry are organized and structured along the lines of the major food ingredients and technologies that are required to better understand their functionalities as well as nutritional impact. This type of setup has been perfected over many years, in some cases such organizations have existed for the better part of 100 years or more. The longevity of food companies in general as well as the underlying business model was discussed in much detail in later chapters. Suffice it to mention here that the original business model of "develop-manufacture-sell" is probably around 1,000 years old, and the food industry always had a development branch, not necessarily called R&D.

More recently, the importance of nutrition grew in the industry and much focus was given to especially health and wellness aspects. This increasing importance of nutrition was difficult to synchronize with ongoing R&D activities that were almost entirely focused on technologies, which served to optimally transform agricultural raw materials into industrial food products. To this day, this "battle" between technology-driven R&D and nutrition-driven R&D is ongoing in many food companies, especially in the larger ones. One important question that comes with this situation is simply this one: who profits most of the typical R&D setup that is found in most food companies? The right answer would obviously be: the consumer. However, it was discussed that the answer is not as simple as this because many more players, especially the company management, are deeply involved. I suggested that top management profits most of the present setup of R&D organizations because they can keep them under close scrutiny and in check. Later chapters in this book suggested alternative R&D setups that might, at least to some degree, reduce this "short-leash" control of management.

12.1.10 People in the food R&D

The other important topic discussed in this context is people, more specifically the type of people that can preferably be found in today's R&D organizations in food companies. I suggested that there are two basic types to be found and what I called: the *hoppers* and the *stayers*. I defined the hoppers as those who have not much patience to remain in any function or any job longer than they think might be necessary, and that they have, mostly according to them, achieved everything that could be achieved during their tenure in a particular function. On the other hand, one can find many stayers, those who are afraid to move on and who always have the feeling that they still have something to discover and add to their present function. These are actually the backbone of every R&D organization because only time and one's own will adds know-how and wisdom. The question of healthy mix and especially personnel turnover was discussed and a percentage of approximately 3 to 5 percent annual renewal was suggested to be a balanced one.

The hoppers are a precious group of employees because they force everyone to always be on alert and on top of his or her function. Overall, I suggested three types of hoppers: super hoppers (strong ego), opportunity hoppers (always open eyes and ears), and "nepo" hoppers (those who have relatives in high places). In the group of stayers, I also suggested three subtypes: the habit stayers (those with strong loyalty), "enthu" stayers (totally enthusiastic about their projects, all projects), and finally the no-perspective stayers.

12.1.11 Discovery and innovation: More projects

The role of discovery and innovation in the context of food R&D was discussed. The often-heard slogan "innovate or perish" may sound exaggerated but is simply true. However, not every food company is really interested in innovative and creative new products and rather prefers what is commonly described by *renovation*, or careful and stepwise renewal of existing products. The fairly well-known relationship between size of company and preference toward either innovation or renovation was briefly discussed. From many years of personal experience and observation, I can safely say that the larger the size of the company, the stronger the trend toward renovation and less toward real innovation.

In this context I mentioned a few project stories that I have personally experienced and that served to illustrate the nature of the following types of projects such as:

- The business briefed project
- The secret project
- The "pet" project
- The never-ending project
- The trial-and-error project
- The please-someone project
- The defensive project
- The knowledge building project

Although these and maybe a few other projects can be found in every food R&D organization, not all of them have really merit and add value to the company.

12.2 UNDERSTANDING INTELLECTUAL PROPERTY

The traditional mantra concerning intellectual property (IP) in the food industry was, and to some degree still is: we want to own everything; it's crucial for the success of our business; it's an important part of the value chain. The various elements that compose IP were discussed in quite some detail as presented here:

- Patents
- Recipes
- Trademarks
- Trade secrets and secrecy agreements
- People: the experts and their actions and results
- Alliances and partnerships

The mantra of wanting to own every type of IP has evolved to a more pragmatic approach, which rather puts emphasis on freely using IP that was jointly developed with external know-how providers. This is particularly important in times when open innovation and innovation partnerships have become largely applied approaches in R&D organizations. To come to an agreement with such an external partner to "freely use" the jointly generated IP, several boundary conditions had to be fulfilled, such as:

- Exclusive and license free usage of results for the food company; the main body of the results remains with the inventor, typically the external partner.
- The usage is defined for a specifically negotiated number and type of products.
- Furthermore, usage is negotiated for a specific geography, for large food companies this could mean world wide.
- Usage is limited to a prenegotiated period of time, to be extended if product success warrants it.

12.2.1 We want to own everything: Should we really?

One of the major reasons why many food companies put so much emphasis on ownable IP is largely based on the belief that IP has a value in its own right. The belief is absolutely correct; however, there is a trade-off between owning everything with all related costs of IP support and the ability to move fast by owning what can be called the *application*, the usage of jointly developed IP within the described boundaries. It is a well-known fact that financial analysts use IP-related information and especially the number of patents applied for and granted per year as an important metric in their company valuations. Hence there is a strong drive by every R&D organization to achieve a large number of granted patents every year.

I discussed patent numbers per year in different industries, and it can safely be said that especially informatics and consumer electronics companies have by far the highest number of granted patents. Although all food companies combined have around 200 patents granted, the chemical industry shows a higher number of around 500; telecommunication is around 2,000; and electronics, computers, and IT services seem to play in a different league with approximately 5,000 patents granted it 2014.

The main question, however, is a different one: what's the real value of patents, the one beyond the valuation by financial analysts? Based on personal experience, I suggested that a typical patent portfolio in the food industry looks as follows:

- Core patents, critical to the success, present and future, of the company: 15–20 percent
- Patents that serve as first line of protection of core, the safety perimeter: 20–30 percent
- Other patents, not directly necessary to protect core and partly unrelated: 50–65 percent

12.2.2 Service: An added value for any food company

The real value of IP lies, however, in the strength of the company's brands in the case of the so-called branded companies. This also holds true for companies and retailers that largely deal with private label and store brands. In addition to the list of elements, which compose IP in a food company, the level and quality of service are of extreme importance and can add tremendous value. IP together with strong brands and elevated service level are the three crucial elements that make the real value of any food company.

12.2.3 What are other companies doing? What is my company working on?

Another important aspect related to IP is its role that it can play in industrial intelligence as well as counterintelligence. What I mean by that is to establish a solid understanding of my own company's IP portfolio, especially knowing and understanding key competencies and technologies. Additionally, by studying as much as possible the IP portfolios of other companies, especially competitors but also seemingly less related ones, can be of tremendous value. This may help guiding the direction of my own work through better understanding the strategic intent of competitors as well as general directions that the industry is heading in.

It was mentioned that listing key capabilities is a preferred pastime of top management, mostly in the disguise of asking the R&D group to establish a list of key technologies. The first answer to this request is not "yes, we will establish such a list" but rather a question to no one in particular, mostly to the experts in R&D, such as "what do we define as key technologies?" Lengthy debates can and do emerge from this question, and I have personally experienced a series of repeated activities linked to this over many years. The typical resulting action consists of cascading the question to the lower echelons. It can easily be expected that every expert who is asked the question of defining key technologies will preferably list his or her typically used technologies as vital.

The list, quite understandably, becomes long and far too complex, and the higher echelons will, also quite understandably, refuse such a long list of "key technologies of the company," with the consequence that the whole exercise is stopped or in the best case, postponed to a later date, maybe to a new set of managers. In this context, it was mentioned that the quest to establish a list of key technologies is quite different from the quest to list key technologies. Although competencies are linked to people, technologies are of more impersonal nature and in theory should be an element with a much longer "shelf life."

12.2.4 I want to know who stands behind the competencies

Linked to the oft-repeated request to list key competencies is the desire to list names alongside the competencies to make it easier to everyone in the company to approach the right people for the right questions and projects. This is a tricky ground because such personalized lists may easily fall in the hands of outsiders recruiters or head hunters, and in the worst case, competitors. This can of course go both ways, and your company might be at the receiving end of such information. I do not suggest any illegal action here at all, but with today's IT possibilities, hackers completely outside the industry might gain access to such information and one never knows what might happen to this highly sensitive and personal set of data. This explains the hesitation to establish such lists, as useful and desirable as they may be.

Another important topic discussed was the approach toward open innovation and innovation partnerships. Food companies still seem to have some anxieties when it comes to opening up to the outside world. To some degree it is understandable because the elements of the value chain in the food industry are of apparent lesser value than in other industries. However, this topic is largely described and discussed in various articles and books, and I have written about this extensively in the past (Traitler, Coleman, and Hofmann 2014).

12.2.5 What's my IP worth?

It may be considered that IP has two types of values: the intangible one that is used in company evaluations by financial analysts and the tangible one that one can discover when trying to sell of or license out any type of IP. This type of activity is not a typical one for food companies, and it is not quite clear why this is the case. There could be several reasons, apart from the general aversion of the food industry toward such an approach. IP in the portfolio of a food company that is unrelated to the core business of the company would be an ideal candidate for out-licensing or selling, yet there are not too many examples. The most likely scenario is simply the lack of value of this type of IP: if it has no value for company A it might simply be without value for company B too, as long as A and B operate in the area of food. Ideally, the search for interested parties would have to go outside the area of food to bring back cash for unused IP. It was suggested that opening up this playing field could become an important additional value creator for food companies.

12.3 NEW APPROACHES AND PERSPECTIVES FOR CHANGE

"R&D in the food industry is inefficient" is said by many. Does this make it any truer? Well, in the eyes of those who allocate company funds to R&D (i.e., management), it seems to hold true, for those who work in R&D it appears absolutely false. As always, the truth lies probably somewhere in the middle. The general belief as expressed by management is simple: R&D is too expensive and rarely delivers on time, on budget, and the expected. These are no surprising statements because they are made by non-experts in the field of R&D. This could be compared to German chancellor Angela Merkel judging the performance of German soccer clubs and giving advice with regard to tactics and players.

However, it is clear that not everything is in order with most R&D organizations in the food industry and real change is required.

12.3.1 Something's wrong in the state of R&D

The question "what is wrong with R&D in food companies" was extensively discussed and analyzed and the following summarizes the list of perceived flaws and possible remedies:

Low perceived output	→ Improve perception by management
Results come too late	→ Increase speed of delivery
R&D is inflexible	→ Provide for change of attitude
R&D is too far detached from business	→ Increase integration with business
No good understanding of business needs	→ Increasingly involve R&D in business decisions
R&D grew too much through acquisitions	→ Reduce overlap and reduce personnel
Perceived asymmetric cost cuttings	→ Fewer, more equitable cost cuttings
Cyclical investment cycle approach	→ Go to anti-cyclical approach
High degree of specialization of personnel	→ Increase in-sourcing of external expertise
Too much redundant personnel	→ Define and keep core expertise in-house; outsource non-core

This is quite a list yet most likely it's not even complete and several other perceived flaws could be added. A rather infamous saying of a former executive of a food company suggested that R&D cost twice what it could cost, however he wouldn't know, which half to cut. This is exactly the dilemma: "which half to cut" or rather which parts of the R&D organization are really excellent and should not be touched and which other elements would require improvement if possible or ultimately would have to be cut. Although budget cuts are a popular means to reduce organizations, especially R&D, the real key to success is innovation; not just any innovation but real relevant and perceivable one. Not renovation, the preferred child in every food company, but innovation, and why not really disruptive innovation?

The ultimate secret to success is a sensible combination of careful cuts, leading to a healthily restrictive work environment that strongly instills the atmosphere of a creative and innovative environment. Sensible budget cuts and constraints can be successful siblings in a field of tension between low spending cuts versus high spending cuts on the one hand and high degree of constraints and low degree of constraints on the other hand.

It is suggested that the optimal area to achieve high creativity paired with high innovation output can be found in the field of low and sensible budget cuts and a fairly high degree of constraints. It must be noted that constraints are not to be confounded with loss of freedom of action, which is an entirely different factor. Constraints are defined as the partial absence of tools, equipment, and resources, encouraging the individual to find alternative solutions and ways around obstacles, thereby increasing the level of creativity often quite substantially.

12.3.2 Consultants: A necessary evil?

The influence of consultants and their attempted guidance was discussed, and it was suggested that consultants are most often brought in to be the amplifier of messages from management, whenever management does not want to convey bad news themselves. On the other hand, properly used consultants can be extremely useful, so there is no one verdict on the role of consultants in any type of organization, including R&D. Whether they are imposed or you have asked for them: make the best out of their presence and possible contributions. Some of the major reasons to bring in consultants were summarized as follows:

- Redirect field of activities
- Pressure from upper management
- Upper management seeks personal glory

There can be many more and it's not always easy to decide when to bring in consultants. There is one element that is common for all consultants: they are not cheap, so one has to really be sure that they can add value.

12.3.3 Lessons from marketing and operations

Another important topic discussed was the role of marketing and especially operations in their attempts to guide and influence R&D to their benefit. This is not an easy topic because historically and for unknown reasons—to me—there is much mistrust between these groups and R&D. There is no one guilty party and all involved need to be criticized for not doing their job properly. Much of this mistrust stems from insecurity, especially insecurity that one's role is not being enough appreciated and recognized. It was mentioned that marketing traditionally has most power in a food company, simply because they have the budgets to support any promising project. However, R&D often feels that they support the wrong projects or even look past R&D and launch renovated products without help from R&D solely relying on potentially faulty consumer research. From personal experience I can confidently say that marketing in general gives orders and makes few compromises. On the other hand, it was also suggested that compromises between two differing positions are probably the worst outcome in business and especially new product development. The solution is simple: R&D needs to be involved in the fact-finding process from the beginning and briefs coming from marketing should be established jointly between marketing and R&D.

The situation is slightly different with operations. Although marketing sometimes can act like a dictator, operations and manufacturing most often act like strict parents who want to teach their kids (i.e., R&D) a lesson, letting R&D know that they are clearly on a lower level. This may sound a bit harsh, but lived experience revealed this reality in many situations. I suggested that one possible explanation for this situation could be the traditionally tense relationship between engineers and scientists.

12.3.4 Evolutionary change in a typical R&D organization

It was suggested that R&D organizations in food companies are inefficient, largely underperforming, and require modifications. Change can and must come in many formats, and such different levels and formats of change were discussed in much detail. The first approach to change, right after denial, is to propose a careful and gradual transition to something better. The next steps that are often taken are to define the targets of betterment (it's trivial but not always the case!) and then suggest a gradual, slow, harmonious transition without hurting the organization too much. The different elements of gradual and careful change were already listed under perceived flaws and possible remedies, so no need to repeat them again. Change, however, requires people to undertake action, and it is suggested that the only group that can actually successfully initiate and carry out change is the R&D group itself, ideally through proactive action. What is meant by proactive action is simple: undertake changes before they are taken for you.

12.3.5 How would consumers see changes in the food industry's R&D?

The food industry is part of an industry that is defined by the term *fast-moving consumer goods*. It's not its only member but probably the most important one. Fast-moving consumer goods create a number of expectations on the part of consumers such as: fairly low cost, always available, convenient to handle, and no side effects. The latter could also be formulated as "not bad for me," better actually as "good for me and my well-being." However, the relationship between consumers and industrially manufactured food is not always and easy and joyful one. Even if the large food companies don't want to see this reality, consumers in many parts of the world become increasingly critical toward industrial food. The reasons are manifold and could be summarized in a consumer wish list when it comes to ideal food: it should be as natural as possible, it should be organically grown (sometimes also called "bio"), it should not lead to allergies or other unpleasant reactions, for some it should not contain lactose, for others no gluten, yet for others no meat and a few more "nots." The laundry list could become long, and ultimately consumers want their own personal food and industrial food fits less and less into this framework of personal desires.

It is clear that it would be extremely complicated for food companies to respond to all these points on the increasingly long wish list and some companies take this as an excuse to paint some or many of these consumer wishes "fads" and believe that by doing so they can wait it out and everything will be OK again. It was suggested that such an approach may have worked in the past, but it certainly works much less effectively today and will be outright wrong in the future. However, food companies still have a hard time to accept this and initiation of change, especially and foremost in their R&D organization. Food companies traditionally are risk averse, and there is no real climate for embracing change, any type of change easily.

12.3.6 Consumer research isn't everything; sometimes it's actually the only thing

The role of consumer research was already mentioned, but it should briefly be reiterated here that carrying out consumer and market research is still one of the preferred pastimes of a food company, especially of course its marketing departments. There is the persistent hope that increasing amounts of market research would lead to an increasing amount of product successes. The almost tragic reality is that pragmatic and continued observation of consumer and market trends in different parts of the world are less believed than results from sophisticated and limited (in size) consumer research. It is well known that depending on the types of questions as well as the type of consumer research one can obtain a set of desired answers.

Cost reductions are one of the major driving forces in every industry, and this holds true for the food industry as well. No surprise here. Consumers are behind this drive because they apparently have been "educated" over time, especially by retailers, that food should actually be cheap and getting cheaper every day. Although food plays an extremely important role in everyone's personal health, consumers don't seem to be prepared to pay what it should cost but rather pay what they believe is the right, low price. Legions are the stories of expensive luxury cars parked in front of the local discounters and their drivers buying the cheapest food products. Something's not right here and change has to come. Bad food leads to bad health, and this costs a lot of money to the public health systems of countries. Again, change has to come, and as suggested, such change should come from the company's R&D organization. And yes, it's not going to be an easy feat! In times when costs are cut, resources become an even rarer commodity and have to be used extremely smartly.

Here a few suggestions for R&D to follow up on that were discussed:

- Turn to fewer and more promising projects.
- Say no to the business more often, yet give them exciting results so the "no" becomes more palatable.

- Anticipate requests better and optimize and streamline project work.
- Increase collaborating with external resources in smart ways (Innovation Partnerships).
- Accept constraints and have smart people find ways around and actually grow.

Every food company will always repeat the mantra that "consumers are in the center," or consumer is king (or queen). It's a nice saying but who is *the* consumer? This question and possible answers are an ongoing point of contention, and there are certainly no simple answers to it. It was suggested that the best answer is probably to take risks and bring the industry forward and pull the consumer with it. This idea is probably most hated by marketing experts who would only accept the other way round, namely that the consumer is the one who pulls, he or she is the one who tells the industry what they want and marketing in turn asks R&D to come up with technical solutions to satisfy the consumers' desires. It was and is suggested that this is probably one of the most outdated concepts in the food industry and should at least be complemented by a more risk-taking attitude and working together with consumers in new and different ways. This will bring a new meaning to "good for the consumers," in which the R&D group of the food company is heavily involved in discovering recognized and well-defined but even more so unrecognized or unmet consumer needs.

The following suggestions were made:

From recognized, met needs	→ Unmet needs
Product and process focus	→ Knowledge and service
Renovation	→ Innovation
Small baby steps/many resources	→ Giant leaps, focused and fewer resources, external know-how
Risk aversion	→ Risk taking (is trivial but needs to be hammered in!

This part of the book also discussed some of the methods that are used today such as video ethnography, and it was suggested that the ideal mix for success is the combination of an appropriate set of observational methods and, almost more important, smart and pragmatic conclusions paired with the ability and desire to take risks. Moreover, the destructive role of too many administrative processes in R&D was briefly discussed. Although compliance-driven branches of any food company, such as manufacturing, require a fairly high amount of auditable administrative roles, R&D can and must get away with a lot less. Such rules are killing two important elements: creativity and risk taking.

12.3.7 Consumer groups and the public opinion

Another topic discussed here was the consumer-driven food R&D as well as the role of consumer groups and the so-called public opinion. Even if the food industry does not seem to have a love affair with consumer groups, it does not mean one has to ignore them. That's the impression that one can occasionally get. Increasingly, consumer groups will strongly interact with the food industry, not in the same ways as in the past when big bad events governed the relationship but increasingly on the basis of simple yet straightforward consumer-interest driven requests by such groups. The industry seems to begin to listen better, yet there is still much room for improvement. It was mentioned that anticipation of issues and new consumer trends and desires should be the name of the game and the industry; R&D organizations especially have to put more emphasis and possibly more resources behind this activity.

This anticipation goes hand in hand with "early warning," which is probably the best approach to getting a handle on possible future problems, which otherwise could just come out of the blue, hitting the company unprepared. To do so it was suggested that:

- Appropriate expert resources are required who are in charge of anticipation (early warning) and smart prevention.
- Ideally, the entirety of the company and not only experts in R&D or manufacturing functions as the extended arms, eyes, and ears of such an early warning system.

It should be recognized that there are unifying elements, bringing together the interests of the consumer as well as the industry and should be the main drivers for the R&D organization. These are:

- Consumer insight
- Everyone's own know-how, expertise, and conclusions based on personal observations
- Public opinion and the interactions with consumer groups.

12.3.8 University perspectives for change

The following discussion was centered on the questions of the academic and university perspectives driving change of R&D organizations in the food industry. One of the important topics that was discussed and analyzed in this context was: why have food science and food engineering developed in parallel to mainstream science disciplines? One of the main reasons is the vertically aligned manner in which the food-related science and engineering activities are organized, whereas science and engineering in general is predominantly structured horizontally. This vertical organization leads to high degree of specialization. This in turn does not allow for easy definition of common features, unless the approach would be highly simplified. This leads to the situation that many experiments have to be planned specifically, thereby almost reinventing the wheel ever so often.

Classical science has taken the diametrically opposite approach and has learned to work with the simplest model system possible to attempt to access the underlying general rules. Food systems in contrast are some of the most complex systems, making such generalist approach almost impossible. This also explains the fairly high costs that are to be expected in food-related scientific or engineering research. This is probably the major reason why the food industry hesitates to sponsor academic research, being well aware that one has to pay a rather high price to achieve comparatively little.

However, despite this, the food industry conducts research, predominantly inside the company, with the clear goal of turning the results of such research into a profit. Because not everything can be achieved through internal expert resources, the world of academia is the first obvious port of call to add complementary resources needed to successfully conduct scientific as and engineering research. Additionally, the objectivity of the results of such research inevitably increases by involving external experts, although a cynical view would be to suggest that because the industry pays, the industry dictates. Those who have been and still are involved in collaborative research with universities know better though. One can try to bend results to better fit in a predefined framework of expectations once, maybe a second time, but ultimately one will be found out and consequences of this can be catastrophic and seriously tainting the image of the food company, if not the industry at large.

12.3.9 IP: The intellectual gold rush

Again, the topic of intellectual property (IP) was discussed and the persistent quest for more IP was compared to an intellectual gold rush. Historically in the years between 1960 and 1970, universities were somewhat idealistic and IP was mainly created for "the good of mankind." This has quite dramatically changed since and most universities around the globe have discovered the importance as well as value of IP, even and especially of such IP as was jointly developed with industry. This is most valid for government or state subsidized universities, for which the University of California is a good example: IP always remains with the university because it belongs to the public who has paid for it through their taxes. This is even true for collaborative research with industry. However, licenses are given to the co-developer at a "better rate." This system has a slight advantage to other systems, for instance the one practiced in the Federal Technical Institutes in Switzerland. In this system the sponsoring industry has the option to pay, in addition to fees for the department's, professor's, and students' works and overhead a certain negotiated amount that gives the sponsor (the industry partner) the right to own the results of this collaboration. The risk here is clearly that one pays money for something that at the end may not even be worth the paper on which the final report was printed.

12.3.10 What does the food industry know about the world of academia?

Perception is everything. Is it really? When employees of a food company, the science and engineering experts, judge academia it is with a view through glasses that are potentially tainted by the science and engineering that surrounds them every day. This can be both a boring and narrow view. However, the following results of such views were discussed and analyzed in some detail:

1. Food scientists, such as lipid specialist, flavor scientist, expert in phase transition, or expert in glass transition.
2. Non-food scientists, such as physicists who work in their "pure" field, mathematicians and so on. It was suggested that all these pure disciplines are largely necessary to do food science, such as the physicist or chemist for carrying out phase transition experiments. This can lead to an artificial and not useful pecking order such as: pure science → applied

science/engineering → food science/food engineering, where food science and engineering are at the believed and perceived lowest level.

3. Engineers—food or otherwise. It was discussed that engineers are real problem solvers, sometimes or even often of problems that are not real.

12.3.11 Nutrition, medical science, claims, and regulatory

Apart from the physico-chemical, the physically and chemically functional side is the entire area of nutritional and health-related functionality that presents a field of activities, both for academia as well as the food industry. Often, this field can become a minefield and related joint research has to be conducted solidly and carefully and with the total willingness to accept whatever results there may be.

12.3.12 Where to get the money from: The role of grants and awards

Governments in most so-called developed countries support and sponsor universities and university research, in some countries like the United Kingdom, even more so if industry is an involved partner in a particular research topic. Often, universities are linked to the idea of a real innovation engine and organizations such as MIT, Stanford, Cambridge, Caltech, and many similar ones are living and certainly thriving examples. Countries such as the United Kingdom have successful government sponsored programs such as Engineering and Physical Sciences Research Council (EPSRC) or Biology and Biological Sciences Research Council (BBSRC); Switzerland has her Swiss National Science Foundation; the European Union has Framework Program; and the United States has a panoply of such programs.

12.3.13 Academics as consultants

Many academics, and especially the more successful and better known ones, increasingly turn to consulting roles and thereby have become important speaking and guiding partners for food companies, especially the R&D groups.

Moreover, such topics as closing the gaps between the different academic disciplines of basic and applied (food) research as well as engineering were discussed. There is not only a need for doing so for efficiency and speed but also for cost and credibility reasons.

12.3.14 What's the future direction?

It is desirable and also likely that a kind of reunification of different science and engineering approaches will happen bringing the food approach and the basic (pure) science approach again together or at least closer to each other.

Other trends that can be observed are the increasing use of research and especially academic research in the area of food as a marketing tool, not quite in the way the detergent industry tries to sell science in their products such as "formula 134 W" or something alike. On the other hand, there is and increasing pressure from the business to try such an approach out in the food industry, as can be seen in the attempts to use science to prove that industrial foods can actually be natural and can help improve health and well-being.

Another trend to be increasingly observed in food science, including disciplines such as nutritional sciences, is crowd sourcing and attempts to involve the anonymous know-how provider somewhere "out there to solve scientific challenges faster and with more supporting substance.

12.3.15 Scientific publication in the future: Multidisciplinary future and collaboration

Publication and dissemination of research especially via peer-reviewed scientific journals has always been the quality indicator of scientific excellence. The trend seems to go away from limited access to scientific literature against a fee toward scientific literature that is publicly accessible. The argumentation behind this is simple: if scientific research is publicly funded, results of such funded research should be publicly accessible. This would of course bring enormous changes and challenges to publishers of scientific journals, and there is no simple answer to this apparent dilemma to date.

It was and is said time and again that real innovation happens at the interface of disciplines. This is quite clearly one of the driving forces behind the stated need for reintegration or reunification of food sciences and the basic—the pure—sciences and especially those, which at first sight are not only not a perfect fit but more so, difficult to understand why they could be useful and necessary to come across the next great discovery in food R&D.

12.3.16 Industry perspectives regarding change in food R&D

The next, rather crucial topic of discussion and analysis dealt with industry perspectives with regard to necessary changes to R&D in the food industry. So-called branded food companies almost always exclusively develop and manufacture their own products under the different families of brands. It is therefore quite logical that any type of R&D work is focused toward these proprietary brands. On the other hand, there are large numbers of private label companies, typically co-manufacturers and co-packers who develop and manufacture for any retailer who is in need of such types of products within their portfolio. There are even cases in which retailers exclusively or almost exclusively manufacture their own store brands, which they then sell in their own stores, such as Aldi in Germany (owner of Trader Joe's in the United States) or Migros in Switzerland.

The food industry is really a champion in complexity and to be able to grasp and define how the food industry at large views and drives necessary changes to their R&D organizations, the following is useful to understand this complexity better. The food industry consists of these elements:

- Agricultural development
- Farming and animal husbandry
- Raw and packaging materials suppliers

- Branded products research and development and their manufacture
- Distribution, distributors, logistics, and retailers
- Private label manufacturers

This list does not include other necessary elements such as finances, legal, regulatory and compliance, quality and safety, and a few more, which can be considered part of the development and manufacture of products at large.

12.3.17 Food and beverage companies are really old

Next, I discussed a bit of history of companies, especially food and beverage companies. The oldest known beverage company is the Weihenstephan brewery in Germany, which was first mentioned as such in documents that date back to 1,050, so almost 1,000 years old. Even today's large food companies such as the Nestlé Company are fairly old; Nestlé commemorates its 150th birthday in 2016. So, it is fair to say that the industry is old and believed to be, because of this longevity, rather successful. This has not held the industry back to try out all kinds of new approaches to business and the creation of SBUs in the early 1990s or the drive toward healthy and nutritious products are just two important examples. Both of these examples have heavily influenced and shaped the type of work that the R&D organizations were supposed to carry out and has, to the detriment of success, even influenced structures and internal organizations of R&D.

A possible company structure of the so-called *complex company* was presented and discussed. The following elements that make up this complex company were suggested as follows:

- SBUs and marketing
- Product-related businesses and marketing
- Markets (countries) with their own marketing and manufacturing
- Regional management with operational responsibility
- Manufacturing and engineering strategy in global headquarters
- R&D, packaging, and design

Some of the readers have certainly realized that parts of this description pretty closely mirror functions only to be found in large food companies.

Another important topic that was analyzed and discussed was linked to the underlying business model, really the basics that one can find in all food companies. It was mentioned that food and beverage companies are old, and so are the business models. At its core, a food company develops, manufactures, and sells its products. The first monks in the Weihenstephan beer brewery did the same around 1,000 years ago: they developed, manufactured, and sold. Nothing new under the sun here. It was argued because the model has always worked and for such a long time already there is no need for change. Maybe or maybe not. This question was extensively discussed throughout this book and especially in the later chapters. It was suggested that for many reasons, such as consumers' increasing requests for more natural food products as just one of them, change has to come. If the business is not capable of changing or reinventing itself because after 1,000 years the identical business model has lost the energy to do so, then R&D would have to jump in the driver's seat and initiate the much needed change.

12.3.18 Anticipate change or be forced to change

The important topic of anticipated change or forced change was extensively discussed at the examples of IBM (anticipated) and Kodak (forced), respectively. The conclusion that anticipating change is the much preferred pathway is easy to understand; however, inertia and risk aversion are big hurdles to this and often lead to forced and imposed changes.

Finally in this discussion on industry perspectives for change in R&D the perceived value of R&D was extensively discussed with the help of two sets of examples: why R&D is bad and conversely, why R&D is great. Here's the list again.

1. R&D is bad because it is too expensive, has the wrong results and output, is too academic, alternatively not enough academic, has wrong resources and talent hired, is too slow, is not responding to real needs of the company, is too far detached from business reality, cannot be trusted, and does simply not understand the business (enough).
2. R&D is great because it creates and develops innovation, has expertise that supports the business, is the trial-and-error expert, adds credibility to the company, adds value as know-how fortress, is knowledgeable in nutrition and health topics, supports manufacturing, has incredible talent pool and supports learning, works with external experts, and helps alleviate the company's tax burden.

Ultimately, if the business side of the company is not ready for substantial changes yet, the R&D group has to take leadership and show most promising and valuable pathways for such change.

12.4 OUTLOOK TO R&D ORGANIZATIONS IN OTHER INDUSTRIES

It is clear that to better understand needs and especially needs for change in an organization it is paramount to look beyond the borders to other types of industries. An entire chapter analyzed and discussed this topic in much detail and gave many examples and compared many situations. There is one important common denominator: food industry or not, R&D will always be led by science and engineering entrepreneurs, by the knowledgeable crowds, by the interaction between various disciplines and industries and by the unusual routes that corporations will choose to take. Historically, the large corporations with big R&D divisions such as ATT, IBM, or Xerox had a de facto monopoly in their fields. Across all industries, this preferred status has dramatically changed, especially with the emergence of open innovation and more recently crowd sourcing.

12.4.1 And the winner in the innovation competition is

Today, any type of ideas and solutions can be drawn from a huge toolbox and new ideas and solutions that were unthinkable 30 or so years ago have become a trivial reality. Competing for "most innovative company" has become more and more popular and the large innovators of today such as Warby Parker (online eye glasses and sun glasses), Apple, Alibaba (platform for trading and communication), and Google were the front runners in 2015. Google is especially active in the R&D space with Google X, Google Research, and Google Ventures. There are many more similar examples, which all demonstrate one important conclusion: innovation companies in almost all business areas all heavily believe and invest in R&D, a fact that is less recognized by the food industry.

And there is more and more revolutionary on the way such as the "lean startup methodology" or R&DIY. Lean startup focuses on shortening product and especially product development cycles by adopting a combination of business-hypothesis driven experimentation in combination with iterative product releases to achieve a status of "validated learning."

R&DIY, which stands for "research and develop it yourself," is a Web platform that allows a large number of so-called international citizens to create and innovate various open source projects together.

12.4.2 The street is your lab

There is an increasing trend toward open source, which can go as far as opening up the good faith usage of a company's IP (such as did Tesla Motors in mid-2014), hoping that many brains can help to improve the product in important ways. The major driver behind Tesla's move was certainly the hope for a common and global technology and logistics platform for electrical cars in general. The open source drive can go as far as to define "the street as your R&D lab." With decreasing resources and investments into the R&D of most companies, especially food companies, the trend goes to become more street-smart and thrive under conditions of constraint. Because "more-for-more" is not a sustainable approach networking, the usage of large partner networks has become the new reality. Based on this approach, Stanford graduates, together with NGOs and hospitals in emerging markets, created a portable infant warmer "Embrace" for premature babies that does not require electricity. Another example is MittiCool, a fridge made entirely out of clay that consumes no electricity and that was co-developed and designed by engaging local community and/or global partners.

12.5 THE VISION FOR THE FUTURE: TESTING THE VISION

12.5.1 The new reality for the food industry's R&D and for the entire food industry

The entire book thus far was building up the argument that R&D in the food industry is not efficient and does not deliver the expected and hoped-for results. However, what if the entire food industry suffered from the same defaults? What if food companies just sit on their 1,000-years-old established business model and try to push out the inevitable—substantial change—to the next generation of managers? I suggested concrete changes, practical and practicable ones, at least so I believe. The existing generation of managers in the business side as well as the R&D side of food companies will certainly reject some or all of them and will denigrate them as unrealistic or too radical. Realizing this, Chapter 11 presented substantial criticism and analysis of all suggested proposals for change. In other words, anticipating unreflected possible rejection, I attempted to dissect and analyze all proposals for change and new business models.

12.5.2 The new suggested business models

Three business models were proposed and all put R&D more into the center of the food company. The following models were suggested:

- Model 2.0
- Model 2.1
- Model 3.0

Model 2.0 consisted of the following three new elements (some might say nothing is new here because they already do it, at least in part):

- Open innovation and innovation partnerships are fully embraced by company and its R&D group.
- R&D sets the tone and steers the company.
- Retailers become real primary target for company, consumers are only indirect targets; moreover, retailers become real partners through intensified collaboration, especially in the field of new product development.

It was stated that the first element of this model, embracing open innovation and innovation partnerships, is still far from being accepted in many food companies, simply for reasons of wanting to hold on to as many parts of the overall value chain, which, in the food industry is of lesser value than in most other industries. Embracing open innovation is the first, important element, which leads to the second one, which is the acceptance of a simple and always overlooked fact: the consumers are not the first target of attention of a food company but we are really talking about retailers being the first port of call for everything that a food company does, unless it sells directly to consumers.

12.5.3 Brand strength is becoming increasingly volatile

It was discussed that selling the company's own products directly to consumers is not the typical business model in the so-called branded food companies but is rather the rule for private label. With the growth of private labels and store brands, branded products will have increasingly difficult times to find their right place on the shelves of retailers. Hence the call for much intensified collaboration with retailers. Real estate for shelf space at retailers for other than store brand products is becoming increasingly expensive and marketing and other general expenses have been continuously growing over the years; they have been growing at rates that are above those of most other divisions in food companies. Thus, the standard answer was for continuous cost reductions at all levels. Although this in itself is not a bad idea, it does not solve the underlying problem of extremely high retailers' listing fees for branded products. The suggested way out and one of the three key elements of the new business model 2.0 is to really get closer to the retailers, and I don't mean that the sales person of the food company goes for a beer with his or her procurement counterpart from the retailer. The "rapprochement" has to be happening at all levels and especially and quite crucially at new product development levels.

It was suggested that store brands become more popular with consumers and some of the reasons are certainly price point and all under one roof family type of products. There are important lessons to be learned for branded companies and model 2.0 suggested exactly this as one of the three elements: collaborate with retailers and create a common interest to get jointly developed new products as efficiently on the shelves and in front of consumers as possible.

The last and maybe most controversial element of all three in model 2.0 is the proposal to bring the R&D group into the driver's seat. For members of other branches of the business, this is most likely a sacrilegious proposal but then it must not be forgotten that those are the ones who have the most to lose should this ever happen. The most important reason for such a transition has to do with a series of consumer fears that can all best be understood, be taken care of, and communicated by members of R&D. Some of the reasons behind this proposal are the following:

- Not clean and unhealthy ingredients
- Long transportation/distribution pathways
- Large carbon dioxide footprint
- GMO—only good for industry
- Not sustainable development
- Too much waste—logistic inefficiencies
- Too much speculation with agricultural raw materials
- Scarcity of water—the need for a "new agriculture" (precision agriculture)
- Climate change impacting agriculture

The suggested model of an R&D-centric food company, therefore has an "office of R&D vision" at the helm of the company and all other (traditional) functions work toward the goals behind the business model 2.0.

12.5.4 We are not there yet

Although business model 2.0 maybe a first possibly step in the right direction to reduce a number of marketing and other general costs and certainly worthwhile to at least think through and apply gradually, it will not help resolve the increasingly critical view of consumers when it comes to industrial food products. It can be seen everywhere, in many TV debates about food, newspaper articles, blogs, social media, and many more that consumers are more and more falling out of love with industrial food. This is a global phenomenon: in the developed world consumers become increasingly critical and in the developing world, consumers could never really afford industrial food products in the first place except for a few basic staples and inexpensive ingredients. The suggested business model 2.1 has at its heart one possible answer, probably the most promising one, to respond to this growing crisis. The core of suggested business model 2.1 consists of the following elements:

- The new company (the "2.1 company") sells know-how rather than products
- Experts are trained to become expert communicators
- Experts, next to their work of gaining more knowledge and improving existing one, are part-time off-site at retailers
- Experts are additionally online to chat with consumers to share their expertise

12.5.5 This change is going to be really tough

It was suggested that a gradual transition happens, still sharing the field of new product development with the new, the friendly retailers, leaving the entire complex of manufacturing to third parties and creating a business model that brings consumers to subscribe to purchasing food and health-related knowledge, ideally personalized as much as possible. A few requirements apart from accepting such substantial change in the first place and that were discussed are the need for a new breed of scientists and engineers who are foremost great communicators and excellent salespersons and a smart approach to involve consumers in individualized and personalized ways into this rather new business model. Some examples were discussed, showing that similar models do and did exist in the health and performance-related parts of the food industry with mixed successes. A rather long list of suggested changes was presented and discussed in some detail. The transformations should go from R&D being a follower to becoming a leader, from hierarchical structures to flat ones, from fixed structures to flowing organizations, from defined projects to creative trial and error, from content that is only defined by the business to content that is defined and driven by R&D, from FTE-based budgets to content- and workload-determined budgets, from formal workgroups and teams to self-forming workgroups defined by real needs, and finally from mostly inside looking and communicating to outward looking and speaking directly to and with consumers.

12.5.6 Testing the hypotheses: First model 2.0

This book ended with a series of critical questions with regard to proposed business models 2.0 and 2.1, both of which require acceptance of rather radical changes in and to the food company and especially its R&D group. The admittedly leading question was: too good to be true or simply wrong? First, a number of reasons why, in the eyes of members of the food company, it is rather impossible to collaborate more closely with retailers:

- Our product is not successful; therefore we need to pay the retailer in order to…
- Get the best shelf spot by retailer.

- It's a tax write-off anyway.
- Retailer is successful, so we are prepared to pay.
- Listing fees also cover shelf management so we don't have to get involved.
- The retailer threatened to withdraw our product, and finally…
- It's the system!

And there was more with regard to the suggested model 2.0:

- In general, the proposed changes have no merit.
- Proposed changes add no value, might even become value destroyers, and…
- Proposed changes are too radical.

12.5.7 What about suggested business model 2.1? Too disruptive and detached from reality?

A list of potential flaws and weak points was discussed, such as:

- Consumers may not be willing to pay "just" for a service such as obtaining food, nutrition, and health-related knowledge.
- New type of experts is required, this might turn out being complex and difficult.
- Training of such new experts may be time-consuming and long to implement.
- What should happen to existing factories and related assets?

A few cases were briefly discussed, all of which are based on personal experience of one of the authors. It could be demonstrated that all of these cases contained some elements of the suggested new models 2.0 and 2.1 and were, at least in part, rather successful, until such time that the "mother company" tried to bend these smaller companies to their traditional business model of develop, manufacture, and sell.

12.5.8 Finally, here yet another business model 3.0 for the R&D in a food company

One of my personal favorites for new and creative set-ups of R&D groups, the Palo Alto Research Center (PARC) was finally discussed, dissected, and analyzed in some detail. The main driving force for the creation of the "new" PARC back in 2002 was imposing constraints, the mother of all constraints being budget cuts. Ultimately PARC remained a fully Xerox owned subsidiary and had a split of 50/50 with regard to projects for the "mother" (Xerox) and projects for third parties. This brought an entirely new spirit to PARC, certainly deplored by some or many, yet embraced by many others as a chance for new opportunities and especially new fields of learning and serving.

The following transformations are required should business model 3.0 be successfully implemented in the R&D groups of food companies:

From projects exclusively for the own company	→ mixed project portfolio
From expertise and IP exclusively internal	→ expertise and IP shared with third parties as appropriate
From tight exclusivity	→ shared results
From closed innovations	→ "invert" open innovation: R&D becomes know-how provider for third parties

Finally, change can happen if fears are overcome and the unknown is replaced by the known. This takes time and much trust in a better, more efficient, and sustainable future. This was the ultimate purpose of this book: to demonstrate and discuss alternative pathways at the example of the R&D organization in food companies. I do hope that you the reader can take a few or maybe even a lot of pieces of learning to your daily grind and make it more exciting and most of all, fun. One thing, however, is certain: change will happen. Make sure that you will be part of it.

So long, until we eat again!

REFERENCE

Traitler, H., Coleman, B., and Hofmann, K. 2014, F*ood Industry Design, Technology and Innovation* Wiley-Blackwell, Hoboken, NJ.

Index

Food Industry R&D: A New Approach, First Edition. Helmut Traitler, Birgit Coleman and Adam Burbidge.